Table of Contents

Part 1 ⇒ Mobile Data Services

▸ **Data Services in GSM** / Overview CSD, SMS, USSD, HSCSD, GPRS, EGPRS, ECSD / Performance / Data Throughput Rates / Applications

▸ **Migration Path from GSM/CSD → EDGE** / Hardware Changes / Provisioning

▸ **Dependency of Network-Architecture on Data Standard**

▸ **The EDGE Family** / EGPRS / ECSD / Compact EDGE / UWC-136

Part 2 ⇒ The Physical Layer of EDGE

▸ **RF Layer** / GMSK / 8-PSK / 3 $\pi/8$ - Offset 8-PSK / Performance / Principles / Functions / Impacts on Power Consumption, Cell Size and Power Control

▸ **Burst Structure** / Burst Types for GSM / GPRS / EDGE

▸ **Structure of the 52-Multiframe** / Description and Function / Radio Blocks / Idle and TA Frames

▸ **Logical Channels** / PDCH's

▸ **Coding Schemes** / CS-1 / MCS-1 – MCS-9 (UL / DL) / Coding and Puncturing Scheme (CPS)

▸ **Data Block Families** / Data Block Family A, B and C

▸ **EDGE User Interface** / Multislot Classes (Multislot Transmission, Timing Constraints) / MS-Class A, B,

Table of Contents

Table of Contents

Part 4 ⇒ The EGPRS Protocol Stack

▸ **RLC/MAC (Radio Link Control / Medium Access Control)** / Differences GPRS ⇔ EGPRS

▸ **LLC (Logical Link Control)** / Function / Acknowledged ⇔ Unacknowledged / Ciphering in EGPRS / Message Format and Coding

▸ **SNDCP (SubNetwork Dependent Convergence Protocol)** / Function / Compression / Message Format and Coding

▸ **Frame Relay / Network Service and BSSGP (Base Station Subsystem (E)GPRS Protocol)** / Introduction to Frame Relay / Overload Defense / BVCI / Function of NS and BSSGP

▸ **GTP (GPRS Tunneling Protocol)** / Function / Message Format and Coding

▸ **Mobility and Session Management in EGPRS** / TLLI / P-TMSI / Routing Area (RA) / Function of RA-Update / READY Timer / GMM- and SM-Message Format / Procedures

---

Part 5 ⇒ Comparison of GPRS ⇔ EDGE ⇔ UMTS

▸ **Performance** / Throughput Rates / Cost / Timely Availability

▸ **Usability of MCS1 – MCS9** / Rural and Urban Environment / Distance from Cell / Implementation Options

---

(1) List of Acronyms

Term	Explanation
3GPP	Third Generation Partnership Project
8-PSK	8 Symbol Phase Shift Keying
AA	Anonymous Access
A-Bit	Acknowledgement Request Bit (LLC ⇔ Logical Link Control)
ABM	Asynchronous Balanced Mode
ACCH	Associated Control Channel
ADM	Asynchronous Disconnected Mode
AGCH	Access Grant Channel
AM	Amplitude Modulation
APN	Access Point Name (⇔ Reference to a GGSN)
ARFCN	Absolute Radio Frequency Channel Number
ARQ	Automatic Repeat Request
AT-Command	Attention-Command
AuC	Authentication Center
BCCH	Broadcast Control Channel

(2) List of Acronyms

Term	Explanation
BG	Border Gateway
BIB	Backward Indicator Bit
BS_CV_MAX	Maximum Countdown Value to be used by the mobile station (⇔ Countdown Procedure)
BSC	Base Station Controller
BSIC	Base Station Identity Code
BSN	Block Sequence Number (⇔ RLC) / Backward Sequence Number (⇔ SS7)
BSS	Base Station Subsystem
BSSAP	Base Station Subsystem Application Part
BSSGP	Base Station System GPRS Protocol
BSSMAP	Base Station Subsystem Mobile Application Part
BTS	Base Transceiver Station
BVCI	BSSGP Virtual Connection Identifier
C/R-Bit	Command / Response Bit
CBCH	Cell Broadcast Channel
CC	Call Control

(3) List of Acronyms

Term	Explanation
CCCH	Common Control Channel
CDMA	Code Division Multiple Access
CDR	Charging / Call Data Record
CEPT	Conférence Européenne des Postes et Télécommunications
CG	Charging Gateway
CGF	Charging Gateway Function
CHAP	Challenge Handshake Authentication Protocol (⇔ PPP)
CPS	Coding and Puncturing Scheme
CS	Coding Scheme
CSD	Circuit Switched Data
CSPDN	Circuit Switched Public Data Network
CS-X	Coding Scheme (1 – 4)
CV	Countdown Value (not Curriculum Vitae ;-)
DCS	Digital Communication System
DHCP	Dynamic Host Configuration Protocol

(4) List of Acronyms

Term	Explanation
DL	Downlink
DLR	Destination Local Reference
DNS	Domain Name System
DPC	Destination Point Code
DRX	Discontinuous Reception
DTAP	Direct Transfer Application Part
DTX	Discontinuous Transmission
ECSD	Enhanced Circuit Switched Data (⇔ HSCSD + EDGE)
EDGE	Enhanced Data Rates for Global Evolution
EGPRS	Enhanced General Packet Radio Service
E-GSM	Extended GSM (GSM 900 in the Extended Band)
EIR	Equipment Identity Register
ERAN	EDGE Radio Access Network
ESN	Electronic Serial Number (North American Market)
ETSI	European Telecommunications Standard Institute

(5) *List of Acronyms*

Term	Explanation
FACCH	Fast Associated Control Channel
FBI	Final Block Indicator
FCCH	Frequency Correction Channel
FCS	Frame Check Sequence (CRC-Check)
FDD	Frequency Division Duplex
FDMA	Frequency Division Multiple Access
FIB	Forward Indicator Bit
FISU	Fill In Signal Unit
FMC	Fixed Mobile Convergence
FN	Frame Number
FR	Fullrate or Frame Relay
FRMR	Frame Reject
FSN	Forward Sequence Number
GEA	GPRS Encryption Algorithm
GGSN	Gateway GPRS Support Node

(6) List of Acronyms

Term	Explanation
GMM	GPRS Mobility Management
G-MSC	Gateway MSC
GMSK	Gaussian Minimum Shift Keying
G-PDU	T-PDU + GTP-Header
GPRS	General Packet Radio Service
GSM	Global System for Mobile Communications
GTP	GPRS Tunneling Protocol
HDLC	High level Data Link Control
HLR	Home Location Register
H-PLMN	Home PLMN
HR	Halfrate
HSCSD	High Speed Circuit Switched Data
HTTP	HyperText Transfer Protocol
I+S	Information + Supervisory
IAM	Initial Address Message (ISUP ⇔ ISDN User Part)

(7) List of Acronyms

Term	Explanation
IETF	Internet Engineering Task Force (www.ietf.org)
IHOSS	Internet Hosted Octet Stream Service
IMEI	International Mobile Equipment Identity
IMSI	International Mobile Subscriber Identity
IMT-2000	International Mobile Telecommunications for the year 2000
IP	Internet Protocol
IPCP	Internet Protocol Control Protocol
IR	Incremental Redundancy ($\Leftrightarrow$ ARQ II)
ISDN	Integrated Services Device Network
ISP	Internet Service Provider
ITU-T	International Telecommunication Union – Telecommunication Sector
L2TP	Layer 2 Tunneling Protocol
LA	Location Area
LAC	Location Area Code
LAI	Location Area Identification

(8) List of Acronyms

Term	Explanation
IETF	Internet Engineering Task Force (www.ietf.org)
IHOSS	Internet Hosted Octet Stream Service
IMEI	International Mobile Equipment Identity
IMSI	International Mobile Subscriber Identity
IMT-2000	International Mobile Telecommunications for the year 2000
IP	Internet Protocol
IPCP	Internet Protocol Control Protocol
IR	Incremental Redundancy ($\Leftrightarrow$ ARQ II)
ISDN	Integrated Services Device Network
ISP	Internet Service Provider
ITU-T	International Telecommunication Union – Telecommunication Sector
L2TP	Layer 2 Tunneling Protocol
LA	Location Area
LAC	Location Area Code
LAI	Location Area Identification

(9) List of Acronyms

Term	Explanation
MRU	Maximum Receive Unit (⇔ PPP)
MS	Mobile Station
MSB	Most Significant Bit
MSC	Mobile Services Switching Center
MSU	Message Signal Unit
MT	Mobile Terminal or Mobile Terminating
MTC	Mobile Terminating Call
MTP	Message Transfer Part
NCC	Network Colour Code
NCP	Network Control Protocol (⇔ PPP)
NI	Network Indicator
N-PDU	Network-Protocol Data Unit (⇔ IP-Packet, X.25-Frame)
NS	Network Service
NSAPI	Network Service Access Point Identifier
NSS	Network Switching Subsystem

(10) List of Acronyms

Term	Explanation
OMC	Operation and Maintenance Center
OPC	Originating Point Code
OSI	Open System Interconnection
OSP	Octet Stream Protocol
P/F-Bit	Polling/Final - Bit
PACCH	Packet Associated Control Channel
PAD	Packet Assembly Disassembly
PAGCH	Packet Access Grant Channel
PAP	Password Authentication Protocol (⇔ PPP)
PBCCH	Packet Broadcast Control Channel
PCCCH	Packet Common Control Channel
PCM	Pulse Code Modulation
PCN	Personal Communication Network
PCS	Personal Communication System
PCU	Packet Control Unit

(11) List of Acronyms

Term	Explanation
PD	Protocol Discriminator
PDCH	Packet Data Channel
PDP	Packet Data Protocol
PDTCH	Packet Data Traffic Channel
PDU	Protocol Data Unit or Packet Data Unit
PLMN	Public Land Mobile Network
PNCH	Packet Notification Channel
POP	Post Office Protocol
PPCH	Packet Paging Channel
PPP	Point-to-Point Protocol
PRACH	Packet Random Access Channel
PS	Puncturing Scheme
PSPDN	Packet Switched Public Data Network
PSTN	Public Switched Telephone Network
PT	Protocol Type ($\Leftrightarrow$ GTP or GTP')

(12) List of Acronyms

Term	Explanation
PTCCH	Packet Timing Advance Control Channel
PTCCH/D	Packet Timing Advance Control Channel / Downlink Direction
PTCCH/U	Packet Timing Advance Control Channel / Uplink Direction
PTM	Point to Multipoint
P-TMSI	Packet TMSI
PTP	Point to Point
QoS	Quality of Service
RA	Routing Area
RAC	Routing Area Code
RACH	Random Access Channel
RAI	Routing Area Identification
RAND	Random Number
REJ	Reject
RFC	Request for Comment (⇔ Internet Standards)
R-GSM	Railways-GSM

late">© INACON GmbH 2001. All rights reserved. Reproduction and/or unauthorized use of this material is prohibited and will be prosecuted to the full extent of German and international laws.

(13) List of Acronyms

Term	Explanation
RLC	Radio Link Control
RNC	Radio Network Controller
RNR	Receive Not Ready
RNS	Radio Network Subsystem
RR	Radio Resource Management
RR	Receive Ready
RRBP	Relative Reserved Block Period
SABM(E)	Set Asynchronous Balanced Mode (Extended)
SACCH	Slow Associated Control Channel
SACCH/MD	SACCH Multislot Downlink (related control channel of TCH/FD)
SAPI	Service Access Point Identifier
SCH	Synchronization Channel
SDCCH	Stand Alone Dedicated Control Channel
SDMA	Space Division Multiple Access
SDU	Service Data Unit

(14) List of Acronyms

Term	Explanation
SGSN	Serving GPRS Support Node
SI	Service Indicator
SIF	Signalling Information Field
SIM	Subscriber Identity Module
SIO	Service Information Octet
SLC	Signaling Link Code
SLR	Source Local Reference
SLS	Signaling Link Selection
SLTA	Signaling Link Test Acknowledge
SLTM	Signaling Link Test Message
SM	Session Management
SMS	Short Message Service
SMSCB	Short Message Services Cell Broadcast
SMS-G-MSC	SMS Gateway MSC (for Short Messages destined to Mobile Station)
SMS-IW-MSC	SMS Interworking MSC (for Short Messages coming from Mobile Station)

(15) List of Acronyms

Term	Explanation
SMTP	Simple Mail Transfer Protocol
SNDCP	Subnetwork Dependent Convergence Protocol
SNMP	Simple Network Management Protocol
SNN	SNDCP N-PDU Number Flag
SN-PDU	Segmented N-PDU (SN-PDU is the payload of SNDCP)
SPC	Signaling Point Code
SRES	Signed Response
SSN	Send Sequence Number
SUERM	Signal Unit Error Rate Monitor
TA	Timing Advance
TAI	Timing Advance Index
TBF	Temporary Block Flow
TCAP	Transaction Capabilities Application Protocol
TCH	Traffic Channel
TCH/FD	Traffic Channel / Fullrate Downlink

(16) List of Acronyms

Term	Explanation
TCP	Transmission Control Protocol
TDD	Time Division Duplex
TDMA	Time Division Multiple Access
TE	Terminal Equipment
TFI	Temporary Flow Identity
TI	Transaction Identifier
TID	Tunnel Identifier
TLLI	Temporary Logical Link Identifier
TLV	Tag / Length / Value Notation
TMSI	Temporary Mobile Subscriber Identity
TQI	Temporary Queuing Identifier
TRAU	Transcoding Rate and Adaption Unit
TS	Timeslot
TSC	Training Sequence Code
UA	Unnumbered Acknowledgement

(17) List of Acronyms

Term	Explanation
UDP	User Datagram Protocol
UI	Unnumbered Information ($\Leftrightarrow$ LAPD) / Unconfirmed Information ($\Leftrightarrow$ LLC)
UL	Uplink
UMTS	Universal Mobile Telecommunication System
USF	Uplink State Flag
UTRAN	UMTS Terrestrial Radio Access Network
UWC	Universal Wireless Convergence (Merge IS-136 with GSM)
VLR	Visitor Location Register
V-PLMN	Visited PLMN
VPN	Virtual Private Network
XID	Exchange Identification
XOR	Exclusive-Or Logical Combination

An Overview
of
Mobile Data Services

Table of Contents

- Data Services in GSM

- Migration Path from GSM/CSD to EDGE

- Dependency of Network Architecture on Data Standard

- The EDGE Family

Data Services in GSM:

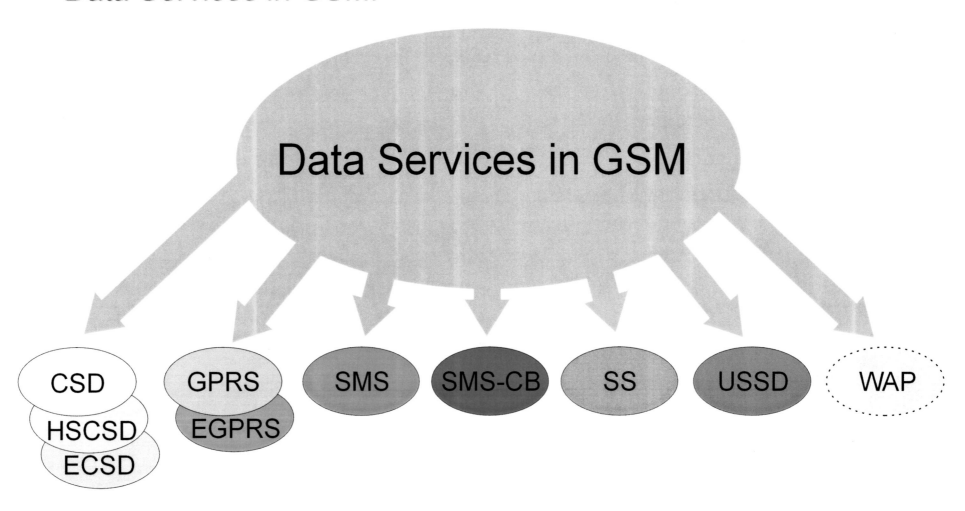

CSD (Circuit Switched Data)

➜ The Circuit Switched Data Service enables data connections to be made between two data terminals within the same or different PLMNs (Public Land Mobile Network) or between a data terminal within a PLMN and a data terminal within a fixed network.

➜ Most services which are provided to fixed network telephone and ISDN users have been included, as far as the limitations of radio transmission permit.

➜ To ensure that the PLMN is compatible with other networks (e.g. PSTN, ISDN, PSPDN, CSPDN) with regard to data services, interworking functions need to be implemented. In essence, these are modems that are implemented as part of the GMSC (Gateway Mobile Switching Center).

➜ CSD (Circuit Switched Data) can produce data rates up to 14.4kbit/s

(1) CSD (Circuit Switched Data):

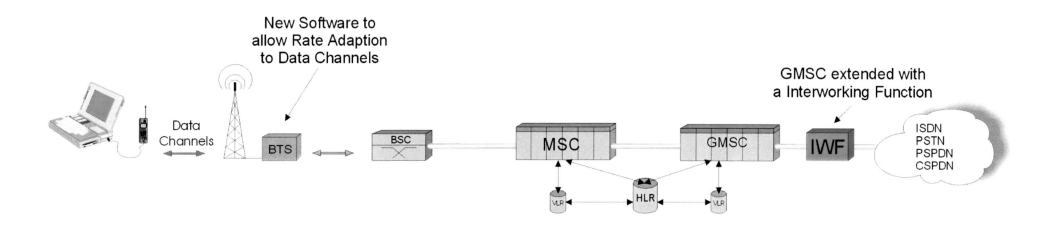

New Software to
allow Rate Adaption
to Data Channels

GMSC extended with
a Interworking Function

Data
Channels

BTS

BSC

MSC

GMSC

IWF

ISDN
PSTN
PSPDN
CSPDN

VLR

HLR

VLR

(2) CSD (Circuit Switched Data):

➜ In CSD, an IWF (Interworking Function) is associated with the Gateway MSC. The IWF enables a PLMN to interwork with fixed networks (ISDN, PSTN and PDNs).

➜ The IWF converts the protocols used in the PLMN to those used in the relevant fixed network. The functions of the IWF depend on the services and the type of fixed network.

➜ The IWF may have no function when the service implementation in the PLMN is directly compatible with that in the fixed network. [3GPP 29.004, 29.005, 29.007 and 09.09.]

(1) HSCSD (High Speed Circuit Switched Data):

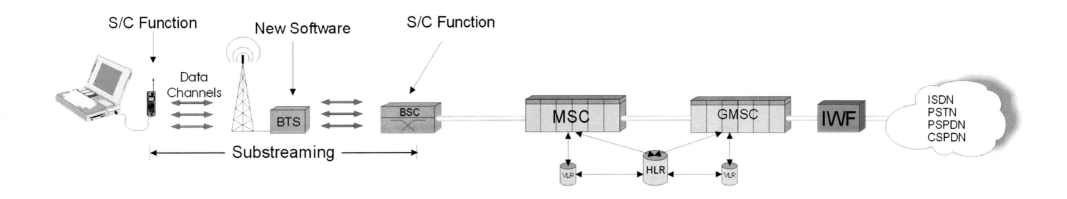

S/C (Split / Combine) Function

(2) HSCSD (High Speed Circuit Switched Data):

→ HSCSD is similar to CSD circuit switched. Resources are permanently allocated and hence cannot be used for other transactions.

→ In HSCSD, the data throughput rate is increased by combining multiple (data) traffic channels for a single connection.

→ HSCSD combines up to 4 data channels on the air interface to provide data rates of up to 57.6kbit/s. However, operators usually offer no more than 38.4kbit/s to their customers in order to limit the complexity of HSCSD since it is regarded as an intermediate technology.

→ The data flow is divided into sub-streams, which require a Split / Combine Function (S/C).

→ The Split / Combine Function is situated both in the mobile station and in the BSC or the TRAU (implementation specific).

(1) ECSD (Enhanced Circuit Switched Data):

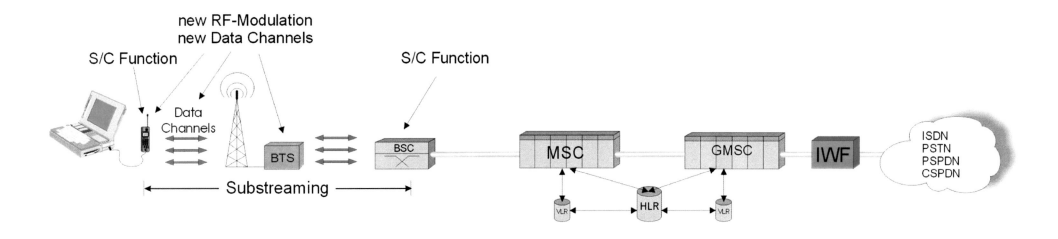

(2) ECSD (Enhanced Circuit Switched Data):

→ HSCSD is enhanced to become ECSD.

→ By implementing a new modulation scheme on the air interface (⇔*The RF-Layer of EDGE),* ECSD achieves higher data rates.

→ Adding a new modulation scheme necessitates an upgrade of the RF hardware in all base stations. New mobile stations are also required.

→ For ECSD, three new Data Channels (E-TCHs) have been defined (28.8kbit/s, 32.0kbit/s, 43.2kbit/s).

→ ECSD supports data rates up to 64kbit/s (Phase 1).

GPRS (General Packet Radio Service)

→   GPRS or General Packet Radio Service is a packet switched technology which is based on GSM.

→   The radio and network resources are only accessed when data actually needs to be transmitted between the mobile user and the network.

→   In comparison with the circuit switched transaction where resources are permanently being accessed, irrespective of whether transmission is actually taking place or not, GPRS saves resources especially in the case of bursty transactions.

→   GPRS enables data to be transferred over IP (Internet Protocol) and PPP (Point-to-Point Protocol). Therefore, all data services which are available on the internet (e.g. FTP, HTTP, E-Mail, Telnet, ...) are also available via GPRS.

→   GPRS also supports the Short Message Service.

(1) GPRS (General Packet Radio Service):

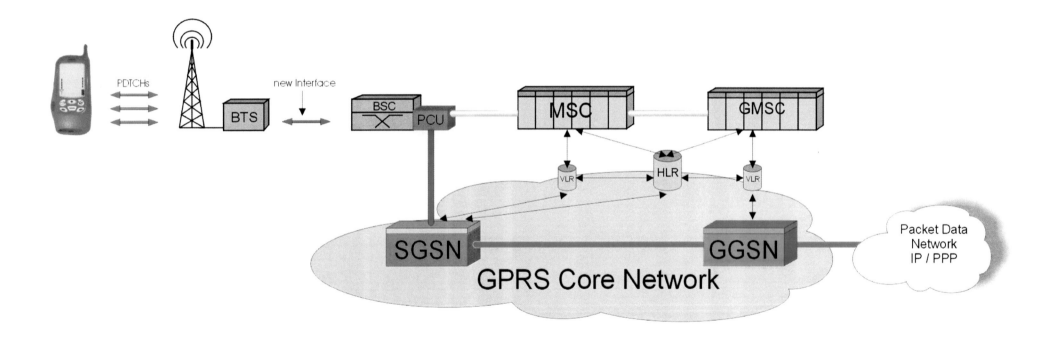

32

(2) GPRS (General Packet Radio Service):

→ In contrast to CSD, GPRS is a packet switched service.

→ In GPRS, Packet Data Traffic Channels (PDTCHs) are defined on the air interface. Depending on the Coding Scheme that is used, a PDTCH may carry up to 21.55kbit/s.

→ This data rate very often requires a new BTS ⇔ BSC Interface which is normally realized in a 4 x 16kbit/s multiplex structure.

→ In GPRS, the BSS is left almost unchanged. The BSC only requires the addition of a Packet Control Unit (PCU), a feature which handles most of the radio related functions of GPRS.

→ For GPRS, an entirely new GPRS Core Network is required.

(1) EGPRS (Enhanced General Packet Radio Service):

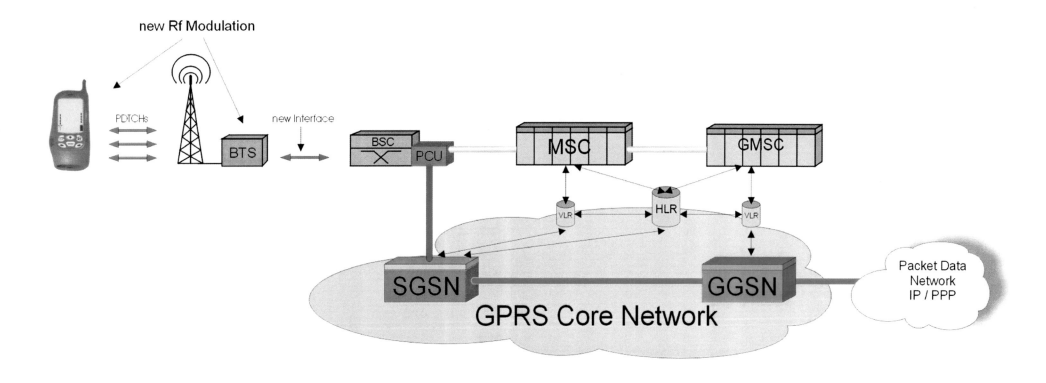

(2) EGPRS (Enhanced General Packet Radio Service):

➜ GPRS is enhanced to become EGPRS.

➜ EGPRS achieves higher data throughput rates than GPRS. This is achieved by introducing a new modulation scheme on the air interface (⇔The RF-Layer of EDGE).

➜ Adding a new modulation scheme requires an upgrade of the RF hardware in all base stations. New mobile stations that support the new RF modulation are also required.

➜ Furthermore, EGPRS introduces new re-transmission procedures and PDTCHs, which must be supported by the mobile station.

➜ Additionally, the BTS ⇔ BSC interface must be modified in order to support the higher data rates.

➜ EGPRS supports data rates up to 59.2kbit/s per time slot.

SMS (Short Message Service):

➔ The Short Message Service provides a means of transferring short messages from one mobile station to another or between an MS and a Short Message Entity such as an internet application (Point-to-Point).

➔ The short message transfer is controlled by a Service Center (SC) which stores the short messages, controls their successful delivery and sends reports back to the mobile station that sent the messages.

➔ A short message has a maximum length of 140 bytes (⇔160 characters) but longer messages are possible by dividing them into multiple short messages.

➔ Short messages may be sent and received during an active call without affecting the call itself.

(1) SMS (Short Message Service):

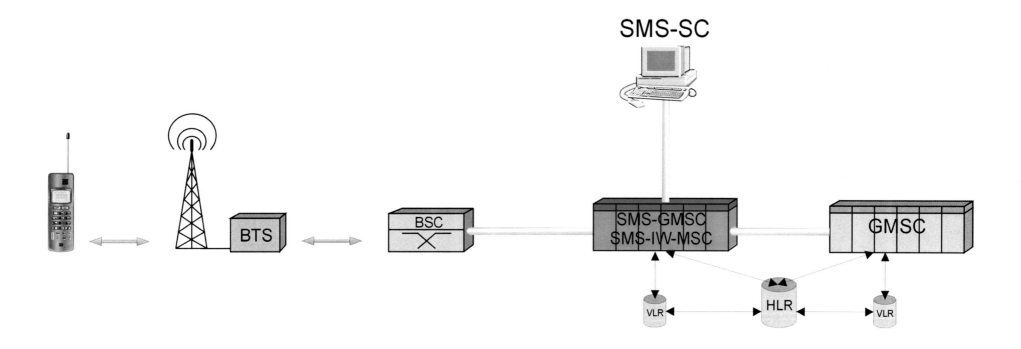

(2) SMS (Short Message Service):

➜ In order to carry out the Short Message Service, an SMS Gateway MSC and an SMS Interworking MSC are added to the GSM Network.

➜ The SMS Gateway MSC is responsible for the transfer of SMS messages from the Short Message Service Center (SC) to the mobile station. The SMS Interworking MSC acts as an interface between the PLMN and a Short Message Service Centre (SC) enabling short messages to be submitted to the SC from Mobile Stations.

➜ The selection of the SMS Interworking MSCs or the SMS Gateway MSCs is at the discretion of the network operator (e.g. all MSCs or some designated MSCs). The functionality of both may be carried out in one MSC.

➜ The Service Center stores the short messages, controls their successful delivery and sends reports back to the mobile station that sent the message.

➜ Short Messages are sent on the radio interface via GSM signaling channels or GPRS PDTCHs. In GSM, in idle mode, short messages are transmitted on SDCCHs, in active mode, the SACCH is used.

SMS-CB (Short Message Service - Cell Broadcast):

→ The Cell Broadcast Service (CBS) consists of cyclical broadcasting of digital information messages to all mobile stations within a given geographical area. (Point-to-Multipoint).

→ SMSCB messages are sent via the BSC to the mobile stations without involving a Short Message Service Center or the Network Switching Subsystem (NSS).

→ SMSCB is an unacknowledged service. As opposed to the SMS, the CBS messages are not acknowledged by the mobile station.

→ The maximum length of an SMSCB message is 82 bytes ($\Leftrightarrow$ 93 characters, the five remaining bits being set to zero).

→ SMSCB messages cannot be received during an active call.

SS (Supplementary Services):

➜ Supplementary Services personalize and enhance the basic services mainly by allowing the user to choose how his mobile terminated calls are to be treated by the network.

➜ Supplementary Service commands initiated by the mobile station are transferred via the network directly to the HLR (Home Location Register) where the requested Supplementary Service is activated.

➜ Supplementary Services are for example *Call Forwarding* and *Call Barring*.

➜ The type of Supplementary Services offered is dependent of the network/operator.

USSD (Unstructured Supplementary Services Data):

➜ Unstructured Supplementary Services Data (USSD) is a means of transmitting non standardized information or instructions on a GSM network.

➜ Unlike SMS, USSD is not a store and forward service but is session oriented. When a user accesses a USSD service, a session is established and the radio connection remains open until the user, application, or time out releases it.

➜ USSD may be used by the operator to offer Supplementary Services to its customers that are not specified in the GSM recommendation. Additionally, USSD is used as a bearer of other services.

➜ USSD text messages can be up to 182 characters in length($\Leftrightarrow$160 bytes, the last six bits being set to zero).

➜ USSD text messages that are received will appear directly in the display of the mobile station.

➜ A mobile station can also initiate USSD operations during an active call.

WAP (Wireless Application Protocol):

→ The Wireless Application Protocol is a standardized procedure enabling a mobile phone to communicate with a server installed in the PLMN.

→ WAP is not a service in itself but it has been defined to provide the consumer with new services. The protocol is designed to economize on the use of the resources that are available in the mobile network.

→ The Wireless Application Protocol runs on top of any underlying bearer. WAP can run on all GSM bearers, Short Message Service (SMS), Unstructured Supplementary Services Data (USSD), Circuit Switched Data (CSD), General Packet Radio Service (GPRS) and Enhanced General Packet Radio Service (EGPRS)).

→ Due to the limitations of mobile stations such as size of display, number of keys etc., WAP applications differ from those used during *normal* internet surfing.

The EDGE Family:

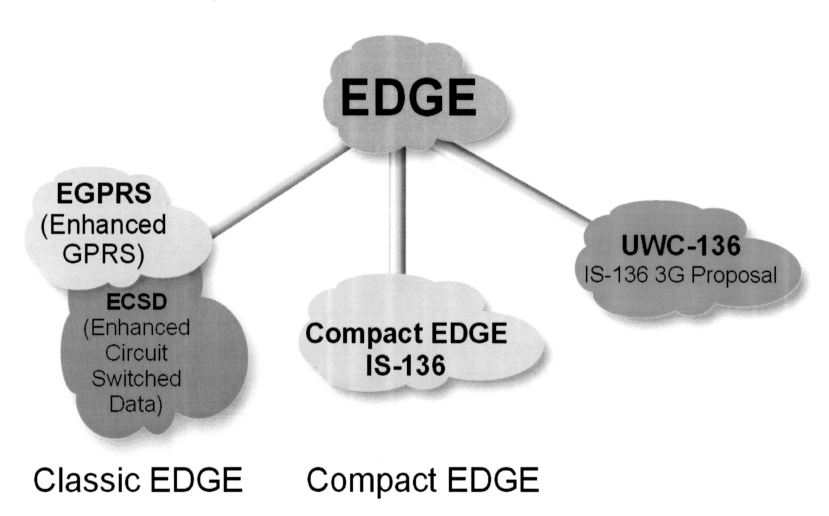

The EDGE Family:

EDGE or Enhanced Data rates for Global Evolution is the name given to the facility that enhances the data rates of the current GSM standards HCSD and GPRS and the US American TDMA (Time Division Multiple Access) standard IS-136.

Therefore EDGE is divided up into different variations:

➔ Classic EDGE

➔ Compact EDGE

➔ UWC-136

Classic EDGE:

➔ Classic EDGE represents a modification of GPRS and HSCSD that enables higher data rates to be attained.

➔ With the implementation of EDGE, HSCSD is modified to become ECSD (Enhanced CSD) and GPRS in improved so as to become EGPRS (Enhanced GPRS).

➔ The higher data rates are realized by means of introducing a new RF- modulation: the 8-PSK Modulation (⇔ *The RF-Layer of EDGE).*

➔ Furthermore, Classic EDGE specifies "intelligent" re-transmission procedures *(⇔ Acknowledged Mode for RLC / MAC Operation),* which permit higher data throughput rates for EGPRS in comparison to GPRS (assuming that the channel quality is identical).

Compact EDGE and UWC-136:

➜ Compact EDGE represents the further development of the US American TDMA standard IS-136 to High Speed Mobile Data. Compact EDGE can be described as an overlay network for IS-136 which is based on the EGPRS standard (⇔ GSM).

➜ UWC-136 is a proposal made by the UWCC (Universal Wireless Convergence Consortium) for moving towards 3G. UWC-136 supports data rates up to 2Mbit/s by means of 8-PSK Modulation.

➜ Indeed UWC-136 would require thousands of man years of development and standardization which no one is prepared to invest. Therefore, it is most probable, that the American operators will also use GSM based standards to pave the way to 3G.

The
Physical Layer
of EGPRS

47

Table of Contents

- The RF Layer of EDGE

- Burst Structures on the Air-Interface

- The 52-Multiframe structure

- Packet Data Channels

- The Modulation and Coding Schemes MCS-1 - MCS-9

- The Data Block Families (A, B, C)

- The User Interface

48

The RF Layer of EDGE:

➔ Two Modulation Schemes are defined in EDGE:

- GMSK (Gaussian Minimum Shift Keying) as in GSM and GPRS
- 8-PSK (8 Symbol Phase Shift Keying)

➔ The chosen RF modulation depends on the quality of the actual radio link.

49

(1) 8-PSK Modulation:

→ In GMSK and 8-PSK the input bit sequence is represented by a phase shift of the RF-signal.

→ In GMSK, a phase shift occurs for each input bit. In 8-PSK however, a sequence of three input bit represents a symbol and the corresponding carrier phase.

→ In 8-PSK, each possible combination of three input bits is allocated to one symbol. Thus 8 symbols (000 to 111) are defined for 8-PSK.

→ Each input symbol generates a phase shift to one of eight defined phase states.

(2) 8-PSK-Modulation:

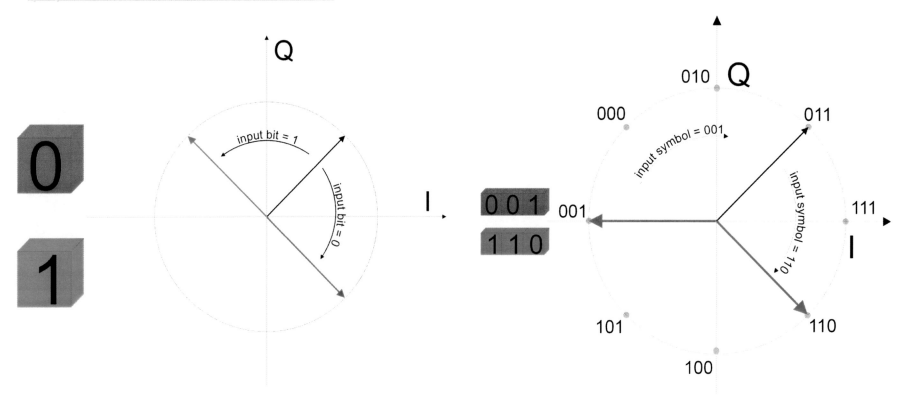

GMSK:
Each input bit causes a +/- 90° phase shift of the RF-vector in the I/Q-plane.

8-PSK:
Each symbol (sequence of three input bits) causes a phase shift of the RF-vector in the I/Q-plane.

(3) 8-PSK-Modulation:

➜ Compared to GMSK, 8-PSK has three times the data throughput capacity. However, its major disadvantage is the inconsistent envelope of the modulated signal.

➜ While the amplitude in GMSK is almost constant, the amplitude in 8-PSK depends on the symbol change of the current modulation.

➜ The next slide illustrates the phase shift for GMSK.

(The following explanation is a simplified illustration of the procedure. Its implementation is vendor specific and there digital filtering as well as oversampling is used to optimize the RF-output signal.)

(4) 8-PSK-Modulation:

For GMSK:

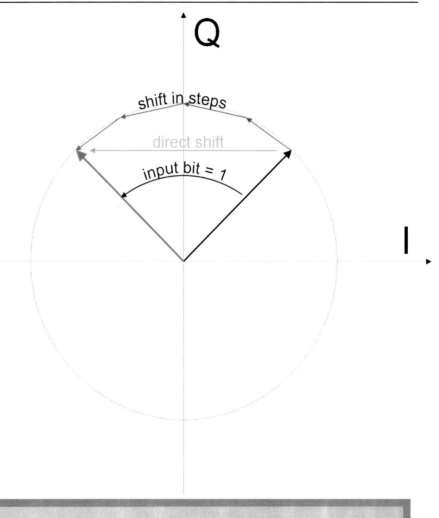

➜ The phase is not shifted directly from e.g. $\pi/4$ to $3\pi/4$. Instead, simply put, the phase is shifted in several steps from one state to the next. With this procedure the amplitude is almost constant.

➜ This procedure may be used because there is always a constant number of steps between two phase states in GMSK.

In the I/Q plane, the amplitude is indicated by the length of the vector.

(5) 8-PSK-Modulation:

For 8-PSK:

→ Due to unpredictable phase changes, the procedure applied to GMSK cannot be implemented in 8-PSK. The number of steps made between two consecutive symbols will usually be different.

→ As a result, the phase must be shifted directly and changes in the amplitude of the RF signal cannot be avoided.

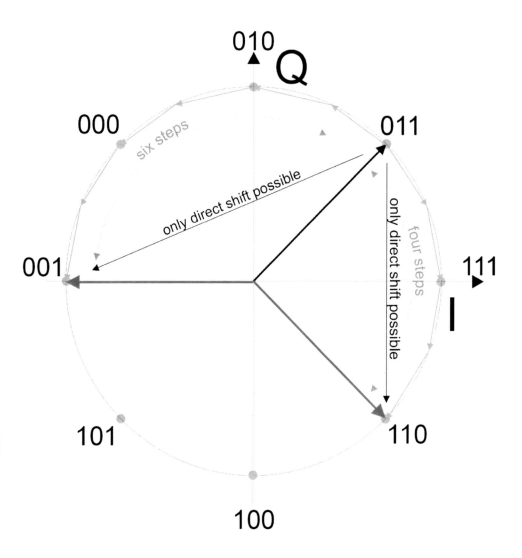

(6) 8-PSK Modulation:

→ The following diagram illustrates the power versus time mask for GMSK. In order to meet the GSM standard, each output burst must fit into this mask. The output power may fluctuate between +/- 1dB of the nominal power.

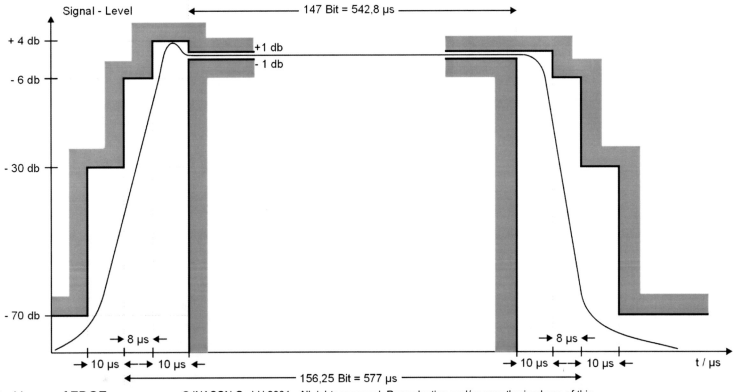

(7) 8-PSK Modulation:

➜ The following diagram illustrates the power versus time mask for 8-PSK. The output power may be -15dB below, minimum due to amplitude degradation, and +4dB, maximum, above the nominal power.

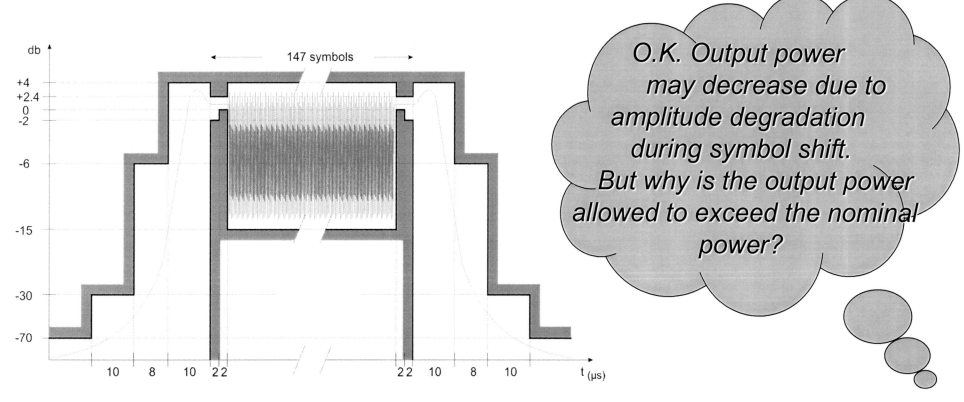

(8) 8-PSK Modulation:

➔ Due to frequency limiting components there is no significant change in phase behaviour which results in slope overshoot of the amplitude value.

➔ The figure illustrates the realistic phase shift between three symbols.

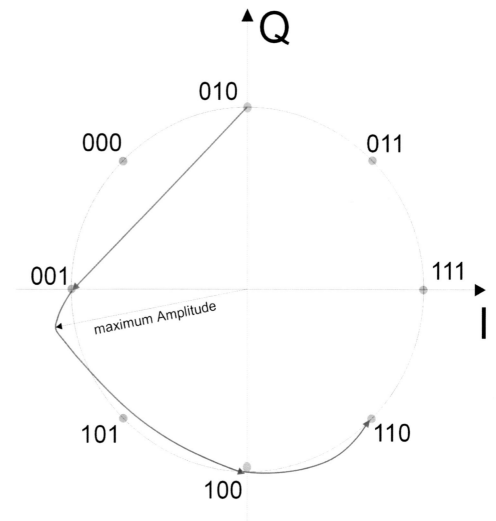

Input bit sequence: 010 001 100 110

57

(9) 8-PSK-Modulation:

→ Another disadvantage of 8-PSK modulation is its greater susceptibilty to error.

→ With increasing interference, the symbols of 8-PSK become more and more diffused. At a certain stage, the receiver is no longer able to distinguish between the different symbols.

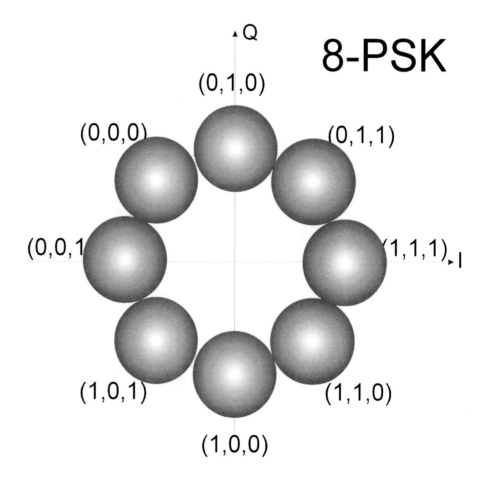

58

(10) 8-PSK Modulation:

Decision space for GMSK and 8-PSK:

→ Clearly, 8-PSK with 8 different symbols is more vulnerable to interference, phase and amplitude degradation, than GMSK.

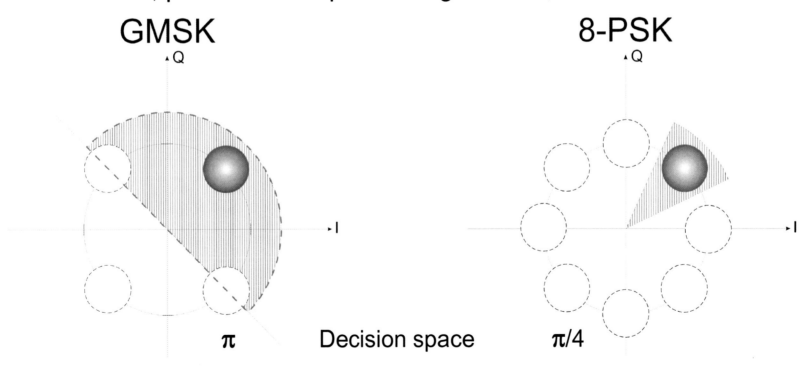

59

(11) 8-PSK Modulation:

→ In order to minimize the bit error rate due to mistakes in transmission, adjacent symbols are arranged on the I/Q plane according to the Gray code.

Gray code:

Between two adjacent symbols only one bit changes.

Hence, if the receiver makes a wrong decision i.e. the symbol falls into the neighbouring decision space, only one bit of the recognized symbol is corrupted.

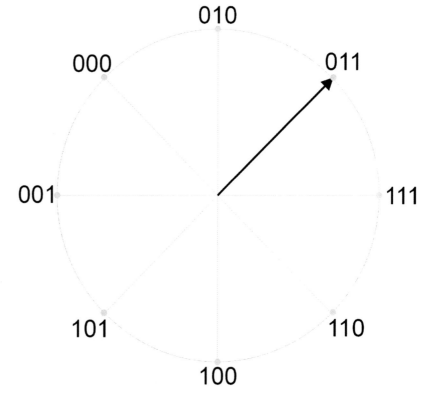

(12) 8-PSK-Modulation:

➔ Furthermore, 8-PSK requires a more linear transmitter characteristic.

➔ Due to the inherent amplitude modulation of 8-PSK, receiver operation in the non-linear area of the characteristic curve must be avoided.

➔ Thus, 8-PSK modulated signals need to be transmitted with a smaller output power than GMSK in order to avoid the respective power amplifier becoming non-linear. This would result in a garbled signal. Consequently, the nominal output power of the transmitters for 8-PSK is lower than for GMSK (⇨cell shrinks).

➔ This effect is also reinforced since 8-PSK is more vulnerable to interference which, in turn, increases with distance.

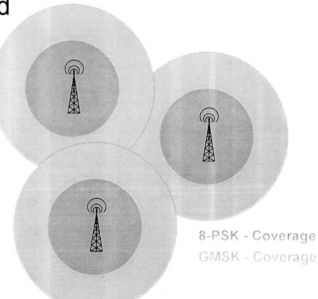

8-PSK - Coverage

GMSK - Coverage

EDGE from A - Z

3π/8 Offset 8-PSK Modulation:

➔ To minimize the amplitude modulation of the RF-signal, EGPRS does not employ plain 8-PSK modulation but 3π/8 offset 8-PSK modulation.

➔ For 3 π/8 offset 8-PSK the symbols in the I/Q plane are rotated by 3π/8 before each phase shift thus preventing the amplitude crossing the zero point.

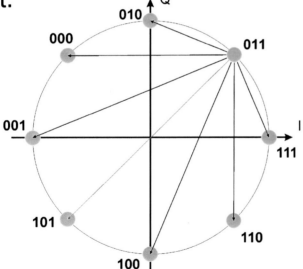

Standard 8-PSK Modulation
Zero Passings (red path)

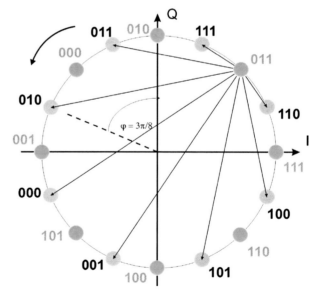

3π/8 Offset 8-PSK Modulation
No Zero Passing

Effect of Offset PSK Modulation:

➜ Phase changes of π result in strong amplitude degradations, amplitude envelope passing zero

➜ 3π/8 rotation avoids passing zero (3π/8 ≠ n x π/4 = Δφ)

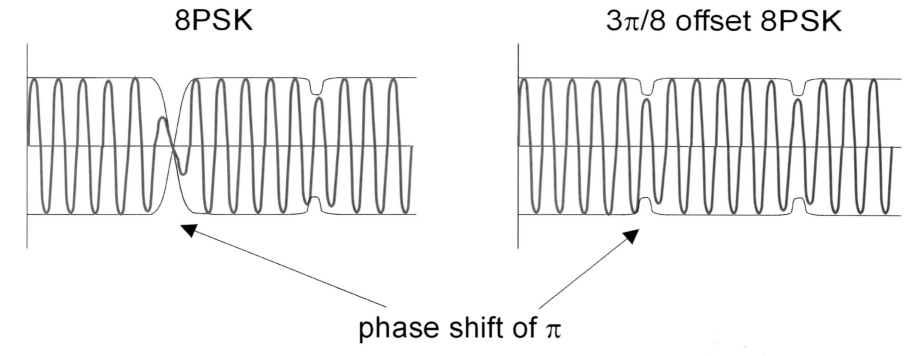

8PSK

3π/8 offset 8PSK

phase shift of π

Blind Detection:

➜ In order to provide compatibility between GPRS and EGPRS mobile stations and to economize on signaling effort, the EGPRS receiver must detect itself if receiving a GMSK or an 8-PSK modulated RF signal (⇨Blind Detection).

➜ For Blind Detection, the MS uses a known bit sequence, training sequences, (⇔Burst Structure for GSM, GPRS and EDGE). The evaluation of the signal samples of this bit sequence which have been received provides information as to whether 8-PSK or GMSK modulation has been used on the air interface.

➜ The implementation of blind detection in the receiver is specific to each vendor.

(1) Burst Structure for GSM, GPRS and EDGE:

➜ GMSK only is used in GSM and GPRS. Therefore, every burst can contain up to 114 bits of useful data.

➜ In GSM, five different types of bursts have been defined:

⇒ The **Frequency Correction Burst (FB)** is the simplest burst of all. It consists of 142 bits, all coded with '0' as well as a head and a tail. The FB is used on the FCCH (Frequency Correction Channel) which has not yet been introduced. The FCCH serves as the beacon of the BTS.

⇒ The **Synchronization Burst (SB)** is used on the Synchronization Channel (SCH) which has not yet been introduced. The SCH conveys the frame number and initial identification information of the cell to the surrounding mobile stations.

Both types of bursts, FB and SB, as well as the respective channels are only applicable on time slot 0 of the carrier that also transmits the BCCH.

(2) Burst Structure for GSM, GPRS and EDGE:

⇒ The **Access Burst (AB)** is used in the uplink direction only when the mobile station does not possess valid information about the current propagation delay of that cell. Therefore, the AB is shortened to ensure that it will fit into the respective receive window of the BTS. This method permits a maximum distance of 35 km between the mobile station and the BTS.

⇒ The **Normal Burst (NB)** is the bearer of almost every kind of information, signaling and payload, in the uplink and downlink directions.

⇒ The **Dummy Burst (DB)** serves a special function on the BCCH carrier where all time slots need to transmit permanently, even when not in use. Hence, all unused time slots of the BCCH transmit dummy bursts. A dummy burst consists of a predefined and fixed bit sequence.
It is necessary to have permanent transmission on all time slots of the BCCH carrier because the BCCH carrier serves as a reference for handover and cell selection decisions.

(3) Burst Structure for GSM, GPRS and EDGE:

Normal Burst:

Synchronization Burst:

Access Burst:

Frequency Correction Burst:

Dummy Burst:

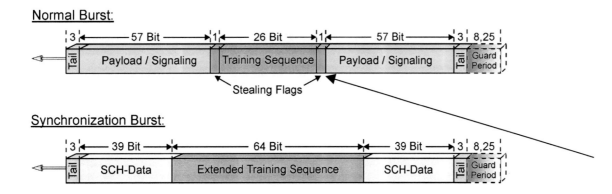

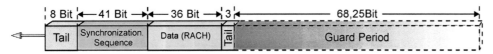

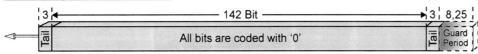

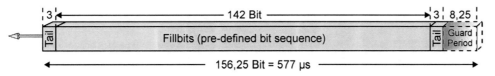

Stealing Flags indicate if and which part of a burst has been "stolen" for FACCH.

(4) Burst Structure for GSM, GPRS and EDGE:

→ In EGPRS two modulation schemes are defined:

GMSK and 8-PSK

→ In GMSK, the same types of burst are used as in GSM.

→ In 8-PSK only the normal burst is defined:

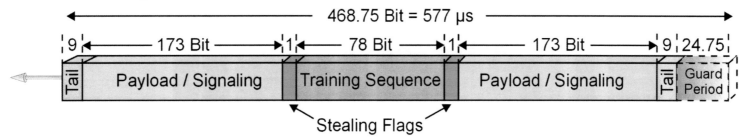

Compared to the 114 bits in a GMSK burst, the 8-PSK burst transmits no less than 346 bits.

Note: Only data on the **PDTCH** can be transmitted via an 8-PSK burst. All signaling is made via the bursts defined for GMSK.

The 52-Multiframe:

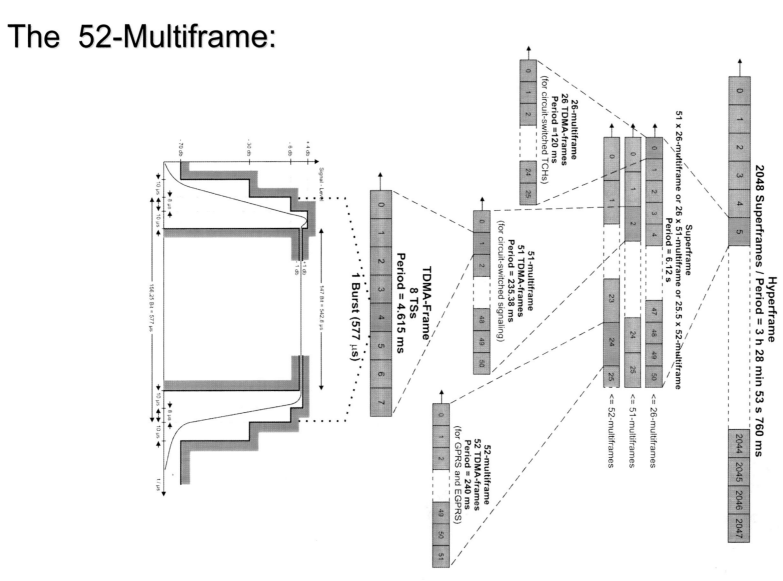

The 52-Multiframe from another Perspective:

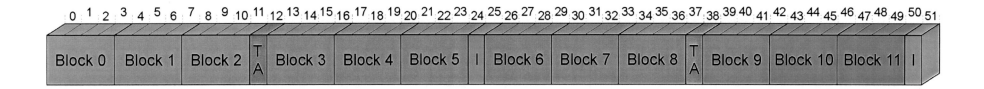

- 12 Radio Blocks for the different Packet Data Channels (PDCH)

- Each Radio Block consists of 4 consecutive appearances of the same time slot within 4 consecutive TDMA-frames.

- Resource allocation in Uplink and Downlink is carried out on the Block Level.

- 2 TDMA frames are reserved for Timing Advance Control (Propagation Delay)

- 2 idle TDMA frames for interference measurements

Comprehension Problems

One 52-multiframe represents 52 times the repetition of one time slot.

This figure represents the use of TS 0.

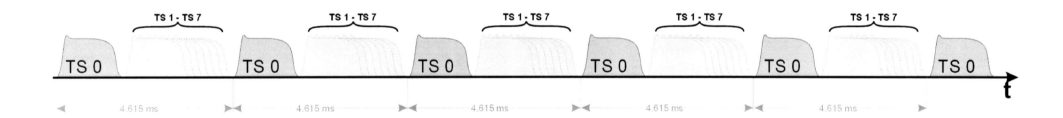

Packet Data Channels (PDCHs) are **Logical** Channels:

➔ PDCHs are configured on one or more time slots of one or more carriers.

➔ PDCHs use the 52-Multiframe structure.

➔ PDCHs can be dynamically allocated by the system.

This also applies for certain types of PDCHs such as the PCCCH or the PBCCH.

➔ PDCH is an overall description of a collection of new logical channels ...

72

Overview:

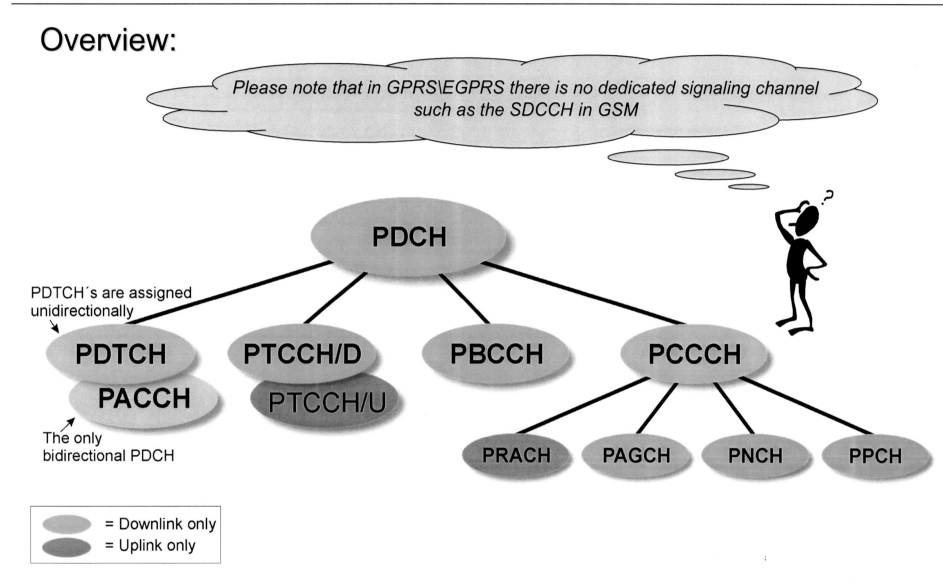

Please note that in GPRS\EGPRS there is no dedicated signaling channel such as the SDCCH in GSM

PDCH

PDTCH's are assigned unidirectionally

PDTCH

PACCH

The only bidirectional PDCH

PTCCH/D

PTCCH/U

PBCCH

PCCCH

PRACH   PAGCH   PNCH   PPCH

= Downlink only
= Uplink only

(1) Names and Functions of the various PDCHs:

→ ## PBCCH (Packet Broadcast Control Channel)

The PBCCH is used to broadcast GPRS\EGPRS related information about a cell to all GPRS\EGPRS enabled mobile stations that are currently camping on that cell. As opposed to the BCCH, the PBCCH can be configured on each time slot of each ARFCN.

→ ## PRACH (Packet Random Access Channel)

Similarly to the RACH in GSM, the PRACH is used to convey the initial network access message from the mobile station to the network (PCU). The PRACH is the only uplink PCCCH.

→ ## PAGCH (Packet Access Grant Channel)

Like the circuit switched AGCH, the PAGCH is used to convey the assignment of dedicated uplink or downlink resources to a mobile station. The PAGCH belongs to the group of PCCCHs.

(2) Names and Functions of the Various PDCHs:

➔ ## PCCCH (Packet Common Control Channel)

PCCCH is an abbreviation for the PRACH, the PAGCH, the PPCH and the PNCH. Note that packet related control information can also be transmitted on the CCCH if no PCCCH is allocated in a cell.

➔ ## PPCH (Packet Paging Channel)

As its name suggests, the PPCH is used to transmit a paging message for GPRS\EGPRS or circuit switched services to the mobile station. Additionally, the PPCH can be used in the GMM state READY to send downlink resource allocation to the mobile station.

➔ ## PNCH (Packet Notification Channel)

The PNCH is used to notify a mobile station about an upcoming Point-to-Multipoint (PTM) transaction.

(3) Names and Functions of the Various PDCHs:

→ ## PDTCH (Packet Data Traffic Channel)

The PDTCH is the bearer for all packet data that is being transferred in the uplink and downlink direction. The GPRS\EGPRS mobile station can transmit and receive on one or more PDTCHs simultaneously. Note that in contrast to the circuit switched TCH, the PDTCH is unidirectional.

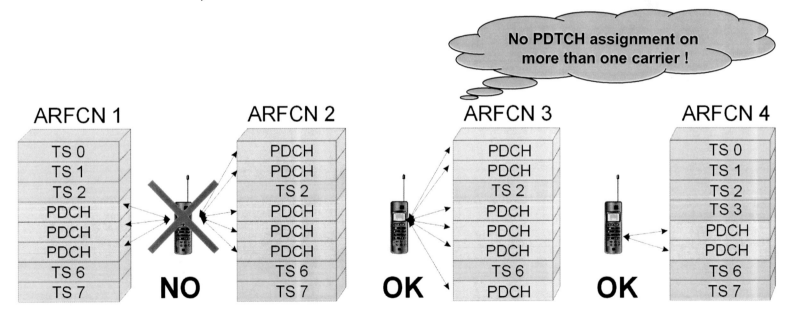

(4) Names and Functions of the Various PDCHs:

→ ## PACCH (Packet Associated Control Channel)

The PACCH is the only PDCH that is available in both directions during a unidirectional GPRS\EGPRS transaction. The PACCH is used to transmit RLC/MAC control information between the PCU and the mobile station.

→ ## PTCCH (Packet Timing Advance Control Channel)

With regard to the PTCCH, a distinction needs to be made between the PTCCH/U (⇐ Uplink) and the PTCCH/D (⇐ Downlink). The PTCCH is only applicable in the 52-multiframe positions 12 and 38 in uplink and downlink direction. This will be addressed in greater detail in part 3.

Coding schemes:

→ In GSM, channel coding is used to protect the information which is to be transmitted against corruption on the radio link.

→ In order to increase data throughput rate in GPRS, four different coding schemes are applied. These coding schemes (CS-1 to CS-4) offer a maximum packet size of 184, 271, 315 and 416 bits.

→ In EGPRS, nine modulation / coding schemes are introduced.

→ However, all three systems employ the coding scheme used in GSM for coding signaling messages. In GPRS, this is CS-1.

(1) Details of CS-1:

→ CS-1 is identical to the coding scheme that is used in GSM for signaling information, on SDCCH, SACCH or FACCH.

→ 176 data bits plus 8 bits for the MAC header are delivered to the encoder.

→ Firstly, the encoder applies the fire coding scheme which adds 40 check bits and 4 tail bits, which are coded with '0000'$_{bin}$.

→ These 228 bits make up the input for the ½ rate convolutional coder which has an output of 456 encoded bits.

→ Since one burst can carry 114 coded bits, 4 bursts are required to transmit one package of 456 bits.

(2) Details of CS-1:

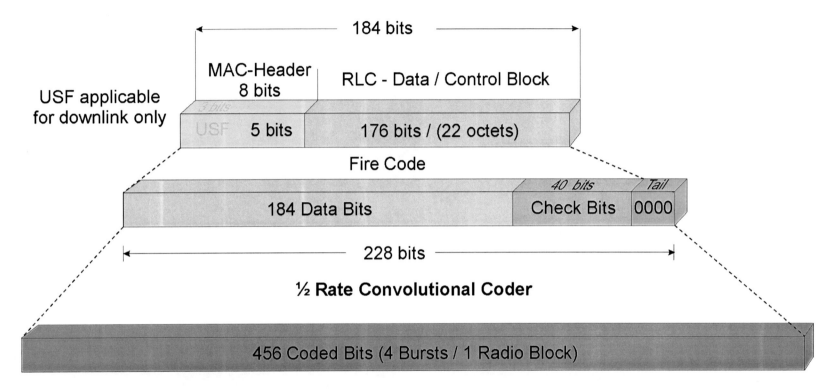

Note: Only CS-1 can be used for the PACCH, PBCCH, PAGCH, PPCH, PNCH and downlink PTCCH

(1) The Coding Schemes MCS-1 to MCS-9 in EGPRS:

➔ In addition to the channel coding schemes used in GSM and GPRS, EGPRS introduces nine new coding schemes (MCS-1 to MCS-9) for EGPRS Radio Blocks (4 bursts / 20ms).

➔ MCS-1 to MCS-4 use the same modulation on the air-interface as in GSM and GPRS (GMSK). A new RF modulation is introduced for MCS-5 to MCS-9: The $3\pi/8$-Offset 8-PSK (⇔*The RF-Layer of EDGE).*

➔ Note that MCS-1 to MCS-9 are only applicable for RLC data blocks, while CS-1 is always used for RLC-control blocks.

➔ For MCS-1 to MCS-9 the block structure differs between the Downlink (DL) and the Uplink (UL) because the header sizes are not the same before channel coding.

➔ In contrast to GPRS, no USF coding is applied in the Uplink direction in E-GPRS.

➔ Each RLC data block contains a two bit RLC Header. Since these 2 bits do not require extra protection, they are encoded along with the data part.

(2) The Coding Schemes 1 - 9 in EGPRS:

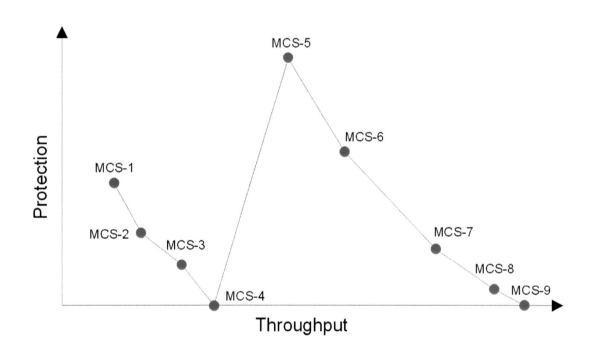

Interdependency between data protection and throughput rate.

Explanation:

➔ EGPRS channel coding is optimized so as to provide very high data transfer rates. In order to save resources on the air interface, puncturing and tail biting is used.

➔ Prior to addressing the channel coding process, these two major processes will be explained first.

(1) Puncturing:

- In GSM and CS-1, the output of the encoder is 456 coded bits. These 456 bits fit exactly into 4 bursts ($\Rightarrow$one Radio Block).

- Note that GSM has not been altered to fit the coded bits from MCS-1 to MCS-9 into 4 bursts only to achieve their higher throughput rates.

- Surprisingly, protection and redundancy are relied upon, added by the 1/3 rate convolutional coder.

- Therefore, at predefined positions, a number of bits are deleted from the output bit sequence. Following that process, only 456 bits for GMSK and 1384 bits for 8-PSK are left for transmission.

- This process is called *Puncturing* and obviously makes MCS-1 to MCS-9 more vulnerable to transmission errors than CS-1.

- On the other hand, puncturing provides a more flexible coding rate (3/4, 2/3, ...) compared with the other option, i.e. changing the coder (1, 1/2, 1/3, ...)

(2) Puncturing:

- EGPRS requires the introduction of different puncturing schemes for puncturing the data block (⇔ *CPS*).

- Two different puncturing schemes (PS 1, PS 2) are used for MCS-1, MCS-2, MCS-5 and MCS-6.

- Three different puncturing schemes (PS 1, PS 2, PS 3) are used for MCS-3, MCS-4, MCS-7, MCS-8 and MCS-9.

- Depending on the MCS used, designated bits are not transmitted for each puncturing scheme (PS 1 to PS 3)   .

- The decoder has the task of recovering the original bit stream from the signal received.

- Obviously, the receiver must know which bits are not transmitted since the punctured bits must be filled before decoding

(1) Tail Biting:

- The illustration below shows a 1/2 convolutional encoder, which delivers two output bits for each input bit. For this encoder, the following applies: each output bit *i* depends on input bit *k* and its sequencing bits k+1, k+3 and k+4.

- For convolutional coding, "tail bits" (mostly 0) are usually added to the tail end of a block in order to initialize the convolutional coder with a defined starting condition for the next data block. Only a defined starting condition enables the data which has been received to be decoded correctly.

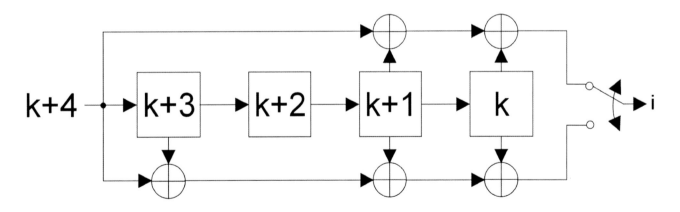

(2) Tail Biting:

- To avoid having extra bits on short blocks (e.g. RLC/MAC Header) and to save resources, tail-biting is used.

- With this method, the encoder is initialized with the last information bits of the datablock. This saves adding the extra bits which are normally added to the tail of the block. The tail is "bitten"off, hence the name tail biting.

- Since the decoder knows which are the last bits of the data block which it has received, it is possible to decode the entire data.

- This method is more expensive and requires more effort from the decoder.

(1) Example of "normal" Convolutional Coding:

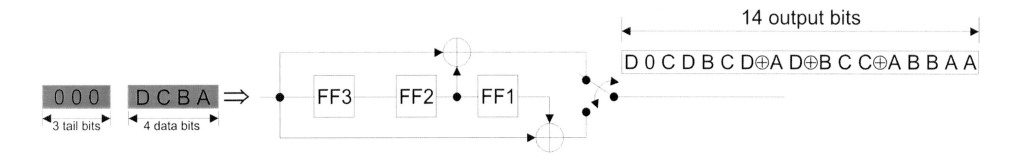

Before the convolutional coding process starts, the flip-flops have a defined starting condition, 0 in this example.

Hence, the first output bits carry information about the input bits e.g. A xor 0 = A.

The tail bits are used to read the last three bits of the data block (B, C, D) and to initialize the flip-flops for the next data block. In this example, the value 0 is used.

Convolutional Coding					
Input	FF3	FF2	FF1	Output	Comment
A	0	0	0	A A	
B	A	0	0	B B	
C	B	A	0	C⊕A C	
D	C	B	A	D⊕B D⊕A	
0	D	C	B	C B	
0	0	D	C	D C	Tail bits
0	0	0	D	0 D	

(2) Example for Tail Biting:

Before the convolutional coding process starts, the flip-flops do **not** have a defined starting condition. The state of the flip-flops correspond to the values of the last three bits of the previously encoded data block.

The output of the first three input bits is discarded since the remaining bits in the encoder (Xn) are not known and the output is therefore undetermined.

Tail Biting					
Input	FF3	FF2	FF1	Output	Comment
B	X3	X2	X1	B⊕X2 B⊕X1	
C	B	X3	X2	C⊕X3 C⊕X2	Discard
D	C	B	X3	D⊕B D⊕X3	
A	D	C	B	A⊕C A⊕B	
B	A	D	C	B⊕D B⊕C	
C	B	A	D	C⊕A C⊕D	
D	C	B	A	D⊕B D⊕A	

Example:

- In the example, the "normal" convolutional coding and "tail biting" convolutional coding is illustrated.

- In the case of normal convolutional coding, three ($\Rightarrow$ number of flip-flops for the convolutional encoder in this example) tail-bits are added for encoding. Firstly, they are used to clear the encoder after the data block has been encoded and secondly, to read the last three bits of the data block which would otherwise remain in the encoder and would not provide any further information.

- In the case of "tail biting" convolutional coding, the encoder is not cleared by the previously encoded data block. Therefore, the last three bits of the data block are added before the data block to initialize the flip-flops of the encoder. The output for these three input bits is discarded since the remaining bits in the encoder (Xn; last three bits of the most recently encoded data block) are unknown which means that the output is undetermined.

- After encoding the data block, the last three bits remain in the encoder (Xn for the sequencing data block). Additionally, they have been used for the initialization of the encoder and are already concatenated with the sequencing input data. Therefore they do not provide any further information and would only waste resources on the air interface.

(1) Details of MCS-1 to MCS-4 Downlink:

ö 178 / 226 / 298 / 354 data bits plus 31 bits RLC/MAC header are delivered to the encoder.

ö The three bit long USF is block coded and extended to twelve bits, the same USF-coding as CS-4)

ö *To guarantee strong header protection, the header part of the radio block is coded independently of its data part*. Eight Parity Bits are calculated for error detection; 1/3 rate convolutional coding and eventually puncturing is used for the correction of errors.

ö In the case of header coding, *tail biting* is implemented to reduce the size of the coded header i.e. no explicit zero value tail bits are added before encoding. The encoder is initialized with the latest information/parity bits,.

ö Four *extra stealing flags* are added in order to attain the correct radio block size of 456 bits. The four *extra stealing flags* in MCS-1 to MCS-4 equal zero.

(2) Details of MCS-1 to MCS-4 Downlink:

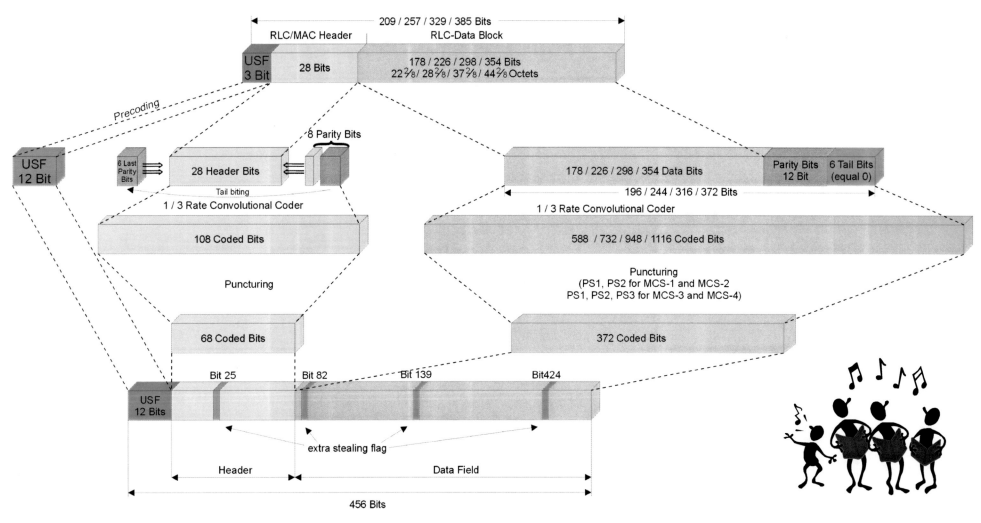

(1) Details of MCS-1 to MCS-4 Uplink :

ö   In the case of MCS-1 to MCS-4 Uplink, the same size of message is delivered to the encoder as for MCS-1 to MCS-4 Downlink.

ö   Since there is no USF in the Uplink direction, there is no USF Coding in the Uplink direction!

ö   To attain the same header sizes after channel coding, 40 header bits are punctured in the Downlink direction and 37 header bits only have to be punctured in Uplink direction.

ö   Additionally, tail biting is used for the Uplink channel coding in order to reduce the size of the header.

(2) Details of MCS-1 to MCS-4 Uplink :

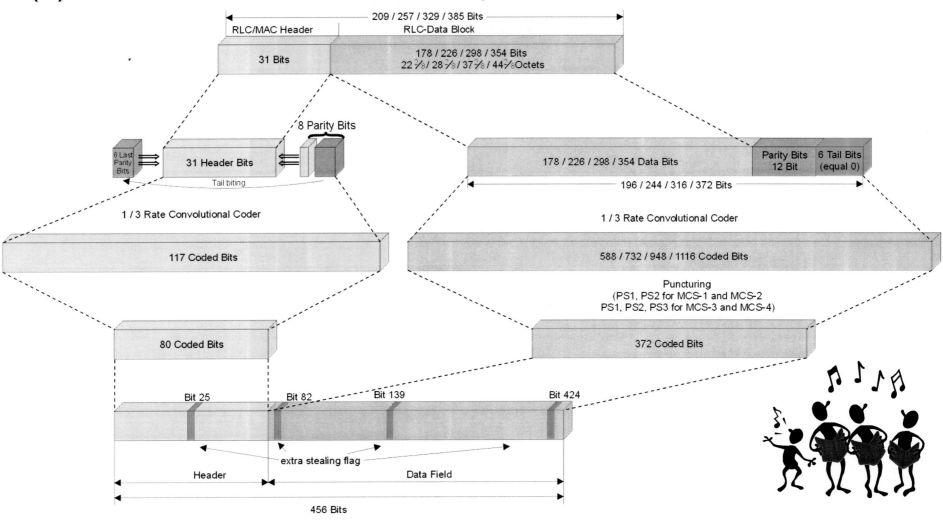

(1) Details of MCS-5 and MCS-6 Downlink :

ö   **MCS-5 to MCS-9 uses 3π/8-Offset 8-PSK on the air-interface.**

ö   450 / 594 RLC data bits plus 28 bit RLC/MAC header are delivered to the channel encoder.

ö   The three bit long USF is block coded and extended to 36 bits.

ö   To guarantee strong header protection, the header part of the radio block is coded independently from the data part. Eight parity bits are calculated for the detection of errors; 1/3 rate convolutional coding is used for error correction. Being a spare bit the last bit (bit 98) is added on to bit position 99.

ö   Tail biting is used for header coding in order to reduce the size of the header.

(2) Details of MCS-5 and MCS-6 Downlink :

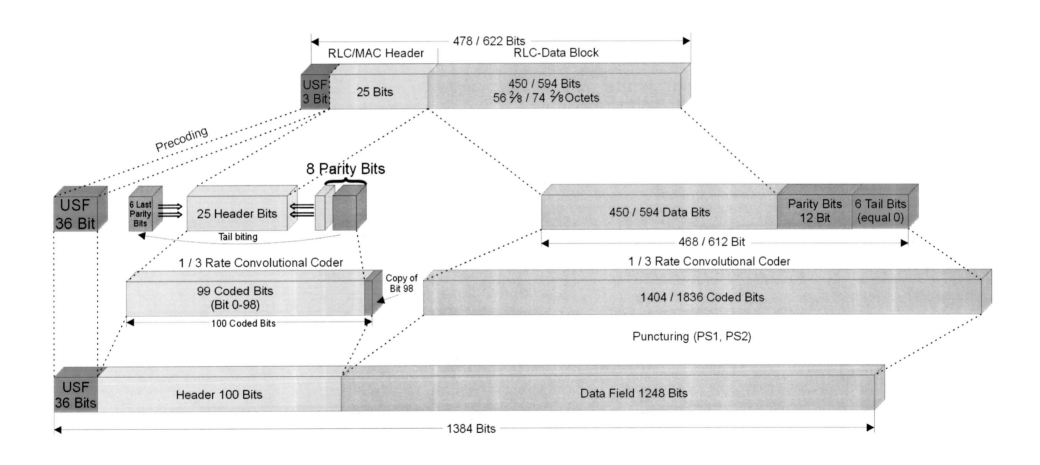

(1) Details of MCS-5 and MCS-6 Uplink :

ö   **For MCS-5 - MCS-9 3π/8-Offset 8-PSK is used on the air-interface.**

ö   450 / 594 RLC data bits plus 37 RLC/MAC header bits are delivered to the encoder.

ö   Being a spare bit, the last bit (bit 134) of the coded header is added on to bit position 135 (see graph).

ö   Because there is no USF in the Uplink, there is no USF coding in the Uplink direction.

(2) Details of MCS-5 and MCS-6 Uplink :

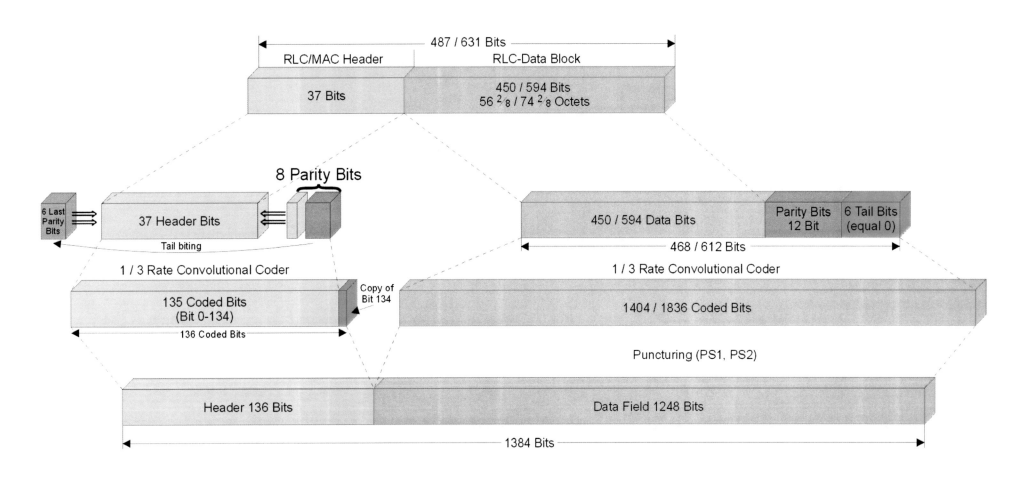

(1) Details of MCS-7, MCS-8 and MCS-9 Downlink:

- **MCS-5 to MCS-9 use 3π/8-Offset 8-PSK on the air-interface.**

- 900 / 1092 / 1188 RLC-data bits plus 40 bit RLC/MAC-header are delivered to the encoder.

- The three bit long USF is block coded and extended to 36 bits.

- For MCS-7 to MCS-9 the payload is transmitted via two RLC data blocks.

(2) Details of MCS-7, MCS-8 and MCS-9 Downlink :

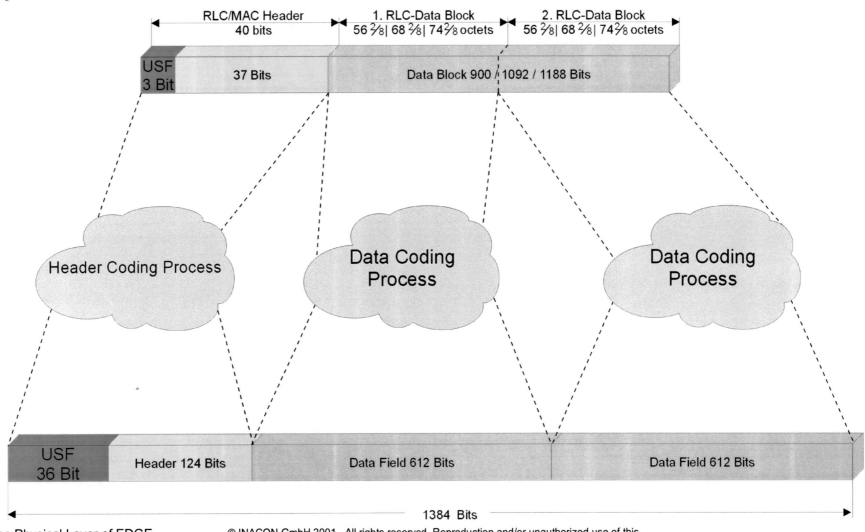

(3) Details of MCS-7, MCS-8 and MCS-9 Downlink :

The Downlink Header Coding Process:

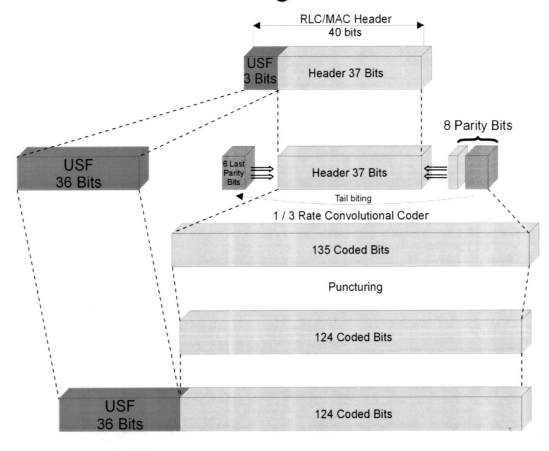

(4) Details of MCS-7, MCS-8 and MCS-9 Downlink :

The Data Coding Process:

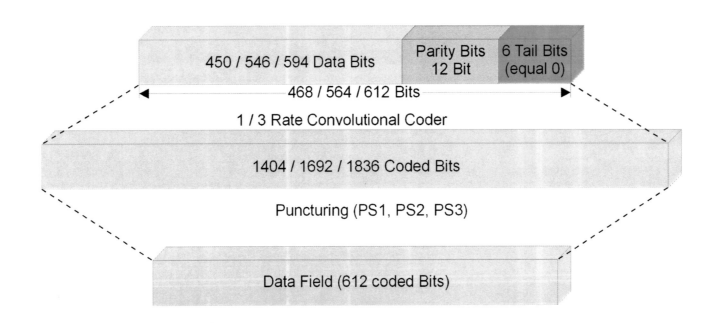

(1) Details of MCS-7, MCS-8 and MCS-9 Uplink:

- **MCS-5 to MCS-9 use $3\pi/8$-Offset 8-PSK on the air-interface.**

- 900 / 1092 / 1188 RLC data bits plus 46 RLC/MAC header bits are delivered to the encoder.

- The Data Coding Process for MCS-7 to MCS-9 Uplink is the same as for MCS-7 to MCS-9 Downlink.

- Since there is no USF in Uplink, there is no USF coding in Uplink direction!!

- For MCS-7 to MCS-9, the payload is transmitted via two RLC data blocks.

(2) Details of MCS-7, MCS-8 and MCS-9 Uplink:

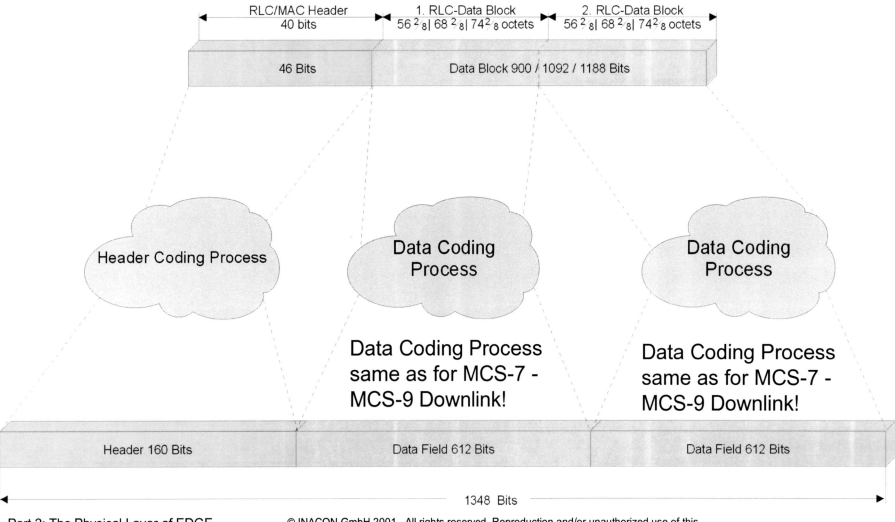

(3) Details of MCS-7, MCS-8 and MCS-9 Uplink:

The Uplink Header Coding Process:

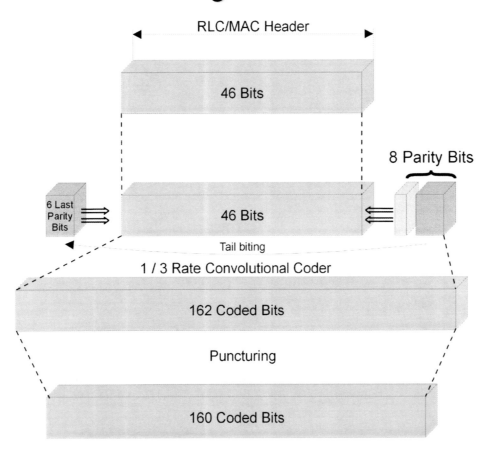

EGPRS Coding Schemes and their Performance:

Coding Scheme	Modulation	Code Rate	Transmission Rate per Timeslot
MCS-1	GMSK	0.53	8.8 kbit/s
MCS-2	GMSK	0.66	11.2 kbit/s
MCS-3	GMSK	0.80	14.8 kbit/s
MCS-4	GMSK	1.0	17.6 kbit/s
MCS-5	8-PSK	0.37	22.4 kbit/s
MCS-6	8-PSK	0.49	29.6 kbit/s
MCS-7	8-PSK	0.76	44.8 kbit/s
MCS-8	8-PSK	0.92	54.4 kbit/s
MCS-9	8-PSK	1.0	59.2 kbit/s

Note: These results only take into consideration the size of the RLC Data Blocks and the fact that a 2bit RLC Header within each RLC Data Block is subtracted.

Discriminating between the nine Coding Schemes:

- Depending on the quality of reception and the error rate, the coding schemes can be dynamically adjusted during a transaction.
Starting with an initial modulation and the coding scheme, dependent on link quality, the actual coding scheme may be changed to MCS-1 to MCS-9 during a transaction.

- Obviously, a method that distinguishes between the coding schemes is required.

- Stealing flags that are not required in their genuine function for GPRS and EGPRS provide the information on the header type within 4 alternating bursts which form one radio block of the 52-multiframe.

- Three different Header types are defined for MCS-1 - MCS-9:
  Header type 1: MCS-7 - MCS-9
  Header type 2: MCS-5 and MCS-6
  Header type 3: MCS-1 - MCS-4
  The header is coded in the same way for all MCS´s which use the same header type.

- The figure on the next slide illustrates the identification of header type 3 (MCS-1 - MCS-4).

The Stealing Flags identify the Headertype:

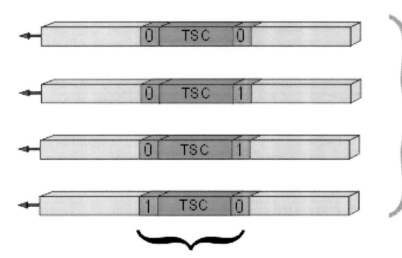

4 alternating bursts, belonging to one radio block

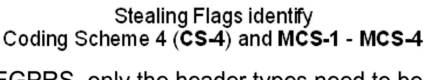

Stealing Flags identify
Coding Scheme 4 (**CS-4**) and **MCS-1** - **MCS-4**

In EGPRS, only the header types need to be distinguished because the header is coded in the same way for all MCS's using this header type.

Each RLC/MAC header contains a *Coding and Puncturing Scheme* (CPS) indication field, which indicates the Coding and the Puncturing of the RLC data block (both blocks for MCS-7 to MCS-9).

Coding and Puncturing Scheme (CPS):

ö In EGPRS, the Coding and Puncturing indicator field is used to indicate the type of Channel Coding (MCS-1 - MCS-9) and Puncturing (PS 1, PS 2, PS 3) used for data blocks.

ö The CPS field is located in the RLC/MAC header.

ö For MCS-1 to MCS-6, where only one RLC data block is transmitted, the CPS field indicates the Coding and Puncturing of the transmitted RLC data block.

ö For MCS-7 to MCS-9, where two RLC data blocks are transmitted within 4 bursts, the CPS field declares the Coding of the two following RLC data blocks. It declares the Puncturing scheme of the first and the second RLC data block separately.

For MCS-7 to MCS-9 both RLC data blocks can be punctured with a

different Puncturing Scheme *(⇔ Acknowledged Mode for RLC / MAC operation)*

(1) Data Block Families A, B and C:

ö   The RLC data blocks in EGPRS are divided into three different Data Block Families (A, B, C). (Note: Family A is split up into Family A and Family A padding.)

ö   These Data Block Families are used for Data Block re-transmission in the acknowledged RLC/MAC operation mode.

ö   Once having been transmitted with a designated MCS, a data block may only be retransmitted with an MCS within the same Data Block Family.

ö   For families A and B, 1, 2 or 4 payload units are transmitted. For family C, only 1 or 2 payload units are transmitted.

ö   When 4 payload units are transmitted (MCS-7, MCS-8 and MCS-9), these are split into two completely separate RLC Data Blocks.

ö   *Data Block Family A padding* is only used for the re-transmission of MCS-8 RLC Data Blocks with either MCS-6 or MCS-3.

ö   In Data Block family A padding for MCS-3 and for MCS-6, the first six octets of each RLC Data Block are padded with zero.

(2) Data Block Families A, B and C:

Family A, B and C:

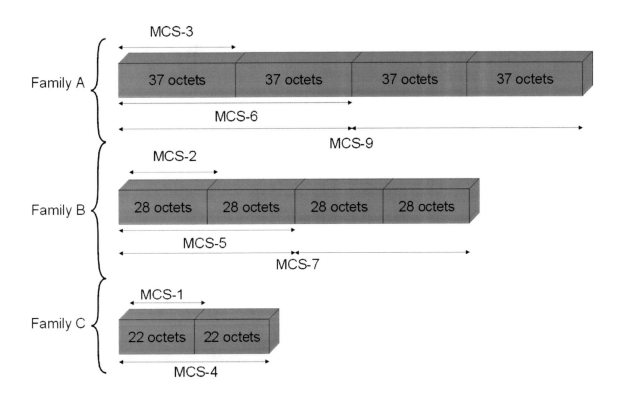

Channel Coding Scheme	RLC data unit size (octets)	Family
MCS-1	22	C
MCS-2	28	B
MCS-3	37	A
MCS-4	44	C
MCS-5	56	B
MCS-6	74	A
MCS-7	2 x 56	B
MCS-8	2 x 68	A
MCS-9	2 x 74	A

(3) Data Block Families A, B and C:

Family A padding:

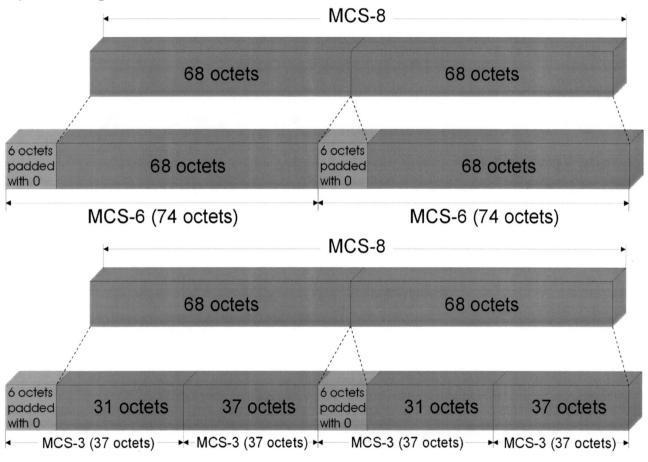

(1) The User Interface:

Multiplexing of GPRS and EGPRS Mobile Stations (MS):

ö The GPRS and EGPRS MSs can be dynamically multiplexed on the same PDCH by using the USF. When uplink resources are allocated to a GPRS mobile, the network must use GMSK, i.e. MCS-1 to MCS-4.

ö Dynamic allocation requires a GPRS MS to read the USF in an EGPRS GMSK block. This is possible by setting the stealing bits in the EGPRS GMSK blocks to CS-4. USF coding and interleaving is carried out as for CS-4.

ö An EGPRS MS cannot differentiate between CS-4 blocks and EGPRS GMSK blocks merely by looking at the stealing bits. This is however not required for USF detection since the USF is signalled in the same way as for CS-4 and MCS-1 to MCS-4.

113

(2) The User Interface:

ö   All coding schemes (CS-1 to CS-4) are mandatory for MSs supporting GPRS. CS-1 is mandatory for a network supporting GPRS.

ö   MSs supporting EGPRS support MCS-1 to MCS-9 on the downlink and MCS-1 to MCS-4 on the uplink. When an MS supporting EGPRS is capable of 8 -PSK on the uplink, it also supports MCS-5 to MCS-9 on the uplink.
A network supporting EGPRS can only support some of the MCSs.

ö   Type II Hybrid ARQ (Incremental Redundancy) is mandatory in EGPRS MS receivers.

(3) The User Interface:

ö   Three different categories have been defined for GPRS and EGPRS:

Class A ø   is a mobile station that can simultaneously perform circuit switched *and* packet switched transactions.

Class B ø   is a mobile station that can monitor circuit switched and packet switched services but cannot operate them simultaneously.

Class C ø   is a mobile station that is limited to monitoring and operating either packet switched or circuit switched services.

115

Details of EGPRS

Table of Contents

- **Network Access Mechanisms**

- **Identification of an ongoing Packet Transaction**

- **MAC Resource Allocation Methods (UL / DL)**

- **Resource Release (UL / DL)**

- **Acknowledged / Unacknowledged RLC Operation mode**

- **Timing Advance Control in EGPRS**

- **Quality of Service in EGPRS**

117

Network Access Mechanisms in EGPRS:

ö   The initial access is used to convey the reason for accessing the network and to determine the distance between the mobile station and the network.

ö   Initial access is carried out on the PRACH if a PCCCH is provided in a cell and on the RACH if the CCCH only is available.

(1) Network Access on RACH:

PACKET CHANNEL REQUEST on RACH:

ö   If the SI 13 (System Information) indicates that a cell is EGPRS capable and that the cell supports EGPRS PACKET CHANNEL REQUEST on RACH, an EGPRS mobile station uses the 11 bit EGPRS PACKET CHANNEL REQUEST messages for one phase access attempts, two phase access attempts and short access attempts.

ö   If the SI 13 indicates that the cell is EGPRS capable and that the EGPRS PACKET CHANNEL REQUEST on RACH is not supported by the cell, the EGPRS mobile station sends the 8 bit CHANNEL REQUEST message and initiates a two phase access request.

(2) Network Access on RACH:

PACKET CHANNEL REQUEST on RACH:

ö   The RACH uses either the CHAN_REQ message which may contain 8 information bits or the EGPRS_PACK_CHAN_ REQ message which may contain 11 information bits, depending on the information in System Information 13 on the BCCH.

ö   Independent of the number of information bits, each (EGPRS_PACK_)CHAN_REQ needs to fit into the shortened access burst (⇔*Access burst)* which can only carry 36 information bits ⇒ *Puncturing*.

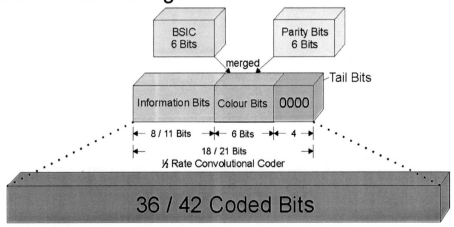

(3) Network Access on the RACH:

How is it possible for the network to distinguish between the 8-bit CHANNEL REQUEST and the 11-bit EGPRS PACKET CHANNEL REQUEST since both messages have both the same format and coding on the Access Burst?

"The 11-bit EGPRS_PACK_CHAN_REQ and the 8-bit CHAN_REQ can be distinguished from each other by different synchronization sequences within the Access Burst!"

Two different *synchronization sequences* for EGPRS ACCESS REQUESTS can be distinguished:

• ACCESS REQUEST with 8-PSK capability in the uplink
• ACCESS REQUEST without 8-PSK capability in the uplink

121

(1) Network Access on PRACH:

PACKET CHANNEL REQUEST on PRACH:

ö   The PACKET CHANNEL REQUEST message has two formats. It can contain either 8 or 11 bits of information. The format, which is to be applied on the PRACH, is controlled by the parameter ACC_BURST_TYPE which is broadcast on PBCCH.

ö   If the 11 bit EGPRS PACKET CHANNEL REQUEST is supported by the cell, PSI1 and PSI13 (Packet System Information), the EGPRS PACKET CHANNEL REQUEST messages are used for one phase access attempts, two phase access attempts and short access attempts.

ö   If the cell can support EGPRS and EGPRS PACKET CHANNEL REQUEST messages are not supported by the cell, the EGPRS mobile station uses the PACKET CHANNEL REQUEST message according to parameter ACC_BURST_TYPE and initiates a two phase access request.

(2) Network Access on RACH:

PACKET CHANNEL REQUEST on PRACH:

ö   On the PRACH, either the PACK_CHAN_REQ message which may contain 8 or 11 information bits or the 11 bit EGPRS_PACK_CHAN_REQ message is used.

ö   Independent of the number of information bits, each (EGPRS_)PACK_CHAN_REQ needs to fit the shortened Access burst (⇔ *Access burst)* which can carry 36 information bits ⇒ *Puncturing.*

ö   It is possible to distinguish between EGPRS_PACK_CHAN_REQ and PACK_CHAN_REQ by the different synchronization sequences within the Access Burst!

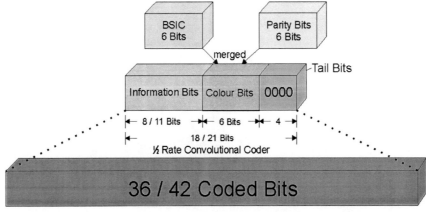

Overview:

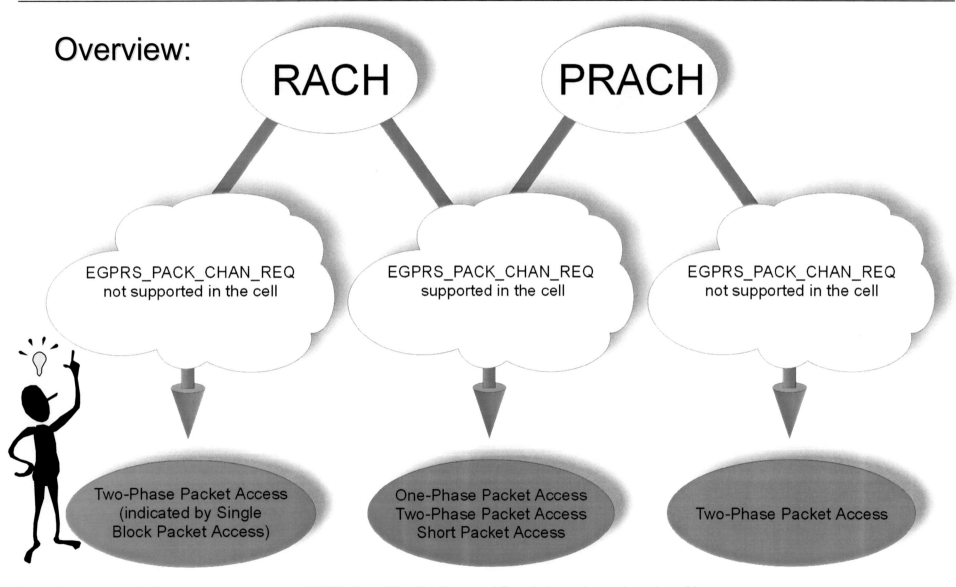

124

(1) Introducing three different Network Access Methods in EGPRS:

One Phase Access:

- Is requested and selected by the mobile station only to acknowledge the RLC operation mode.

- The initial access message is responded to by a suitable resource allocation in the uplink direction.

- Even if the mobile station requests one phase access, the network may still enforce two phase access by allocating a single block only to the mobile station making the request in order to specify its request further.

- *One Phase Access is only possible if EGPRS_PACK_CHAN_REQ is supported by the cell.*

One Phase Access and Contention Resolution:

ö   The initial access message cannot uniquely identify the
    sending mobile station.

ö   Ambiguities cannot be avoided.

ö   Different mobile stations can assume that the same resource
    allocation is destined for it and start using it.

ö   Contention resolution is required.

The Contention Resolution Procedure at One Phase Access:

- *One Phase Access is only possible if EGPRS_PACK_CHAN_REQ is supported in the cell.*

- *The network will assign sufficient resources.*

- *The mobile station includes its identification (⇔ TLLI) in the first uplink blocks that are sent (max. N3104_MAX times). If MCS-7 to MCS-9 are used, both RLC Data Blocks contain the TLLI.*

- *The network replies with a PACK_UL_ACK that includes the TLLI of the addressed mobile station as soon as it can.*

- *If there is a second mobile station using the same uplink resource, it must stop its transmission immediately.*

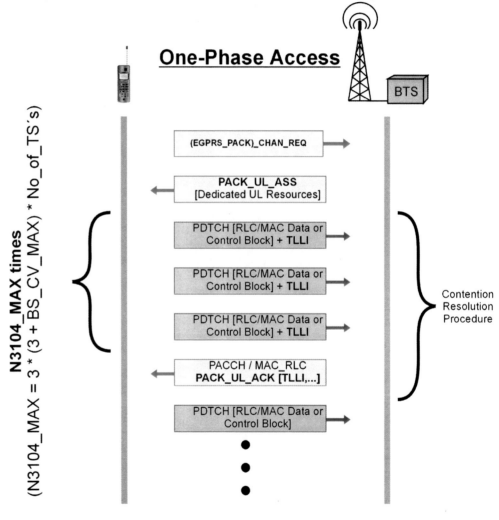

(2) Introducing three different Network Access Methods in EGPRS:

Short Access:

- Is requested and selected by the mobile station.

- If the requested RLC mode is the acknowledged mode and the amount of data n fits into 8 or less RLC/MAC blocks, the mobile station declares Short Access to be the access type.

- Short Packet Access is only possible if EGPRS_PACK_CHAN_REQ is supported in the cell.

- *The number of blocks is calculated on the assumption that the channel coding scheme CS-1 for standard GPRS TBFs, and MCS-1 for EGPRS TBFs are used.*

Short Access:

- *Short Access is only possible if EGPRS_PACK_CHAN_REQ is supported by the cell.*

- *The network assigns as many blocks as are ordered within the CHAN_REQ.*

- *Since no MS-Multislot capability is transmitted within the Short Access CHAN_REQ, only one TS can be assigned for uplink.*

- *The same Contention Resolution procedure as described for One Phase Packet Access is used in order to avoid ambiguities.*

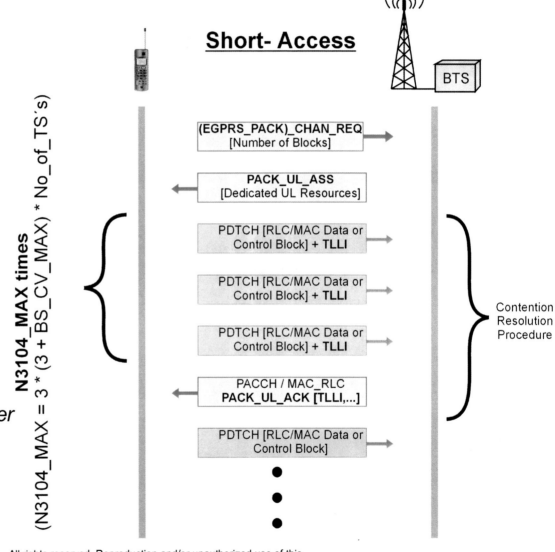

(3) Introducing three different Network Access Methods in EGPRS:

3) Two Phase Access:

- Is selected by the mobile station by asking for single block allocation only (GPRS / EGPRS <=> on RACH) or by explicitly requesting Two Phase Packet Access (GPRS <=> on PRACH; EGPRS <=> on PRACH, or on RACH if EGPRS_PACK_CHAN_REQ is supported)

- Two Phase packet access is mandatory when the unacknowledged RLC/MAC transmission mode is used.

- *In order to inform the network about the EGPRS capability of the MS, Two Phase packet access is also mandatory when EGPRS_PACK_CHAN_REQUEST is not supported by the cell.*

Two Phase Access:

- *Two Phase Access* can be indicated on the following:
  - on RACH or PRACH via EGPRS_PACK_CHAN_ REQ
  - on PRACH via PACK_CHAN_REQ
  - on RACH via CHAN_REQ by ordering a single block only.

- When *EGPRS_PACK_CHAN_REQ is not supported by the cell, the network only assigns a single block to the MS.*

- When *EGPRS_PACK_CHAN_ REQ is supported by the cell, the network can assign one or two blocks to the MS.*

- When *two blocks have been assigned, the MS sends an ADDITIONAL MS RADIO ACCESS CAPABILITIES message.*

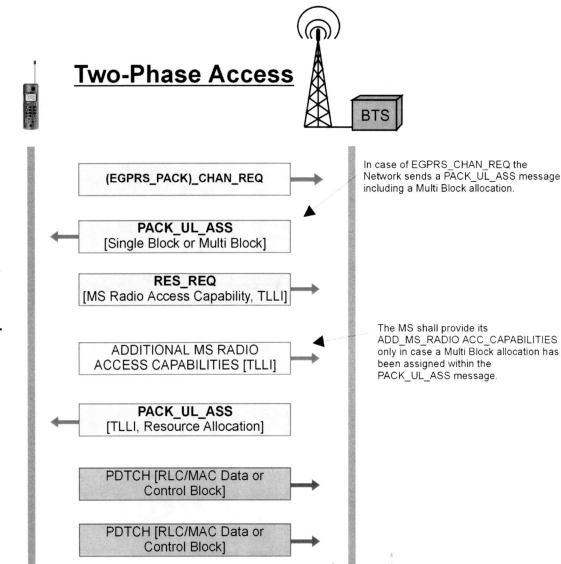

Two-Phase Access

BTS

(EGPRS_PACK)_CHAN_REQ

In case of EGPRS_CHAN_REQ the Network sends a PACK_UL_ASS message including a Multi Block allocation.

PACK_UL_ASS
[Single Block or Multi Block]

RES_REQ
[MS Radio Access Capability, TLLI]

ADDITIONAL MS RADIO
ACCESS CAPABILITIES [TLLI]

The MS shall provide its ADD_MS_RADIO ACC_CAPABILITIES only in case a Multi Block allocation has been assigned within the PACK_UL_ASS message.

PACK_UL_ASS
[TLLI, Resource Allocation]

PDTCH [RLC/MAC Data or Control Block]

PDTCH [RLC/MAC Data or Control Block]

How to Distinguish Ongoing Packet Data Transactions

Problems:

ö   Packet Data Transactions are unidirectional.

ö   Each Packet Data Transaction can involve more than one time slot.

ö   Multiple mobile stations use the same physical resource which is a given time slot, being assigned as PDCH.

ö   Distinction is required for the uplink and downlink directions.

Introducing the Temporary Block Flow (TBF):

ö   The TBF is similar to a channel number in circuit switched transactions.

ö   The TBF or rather its identifier, the Temporary Flow Identity (TFI), identifies an ongoing packet data transfer in the uplink or downlink directions.

ö   The TFI is part of each block being sent in the uplink and downlink directions.

ö   The TFI is unformatted and has a length of 5 bits.

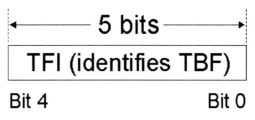

Uplink and Downlink TBFs are independent:

- TBFs are assigned dynamically by the network (PCU)

- It is possible that TBFs with identical numbers are present in the uplink and downlink directions.

- Each TFI is unique for the PDCH's that are allocated to a transaction (and per direction).

- The life span of a TBF is limited to the life span of the related packet data transaction

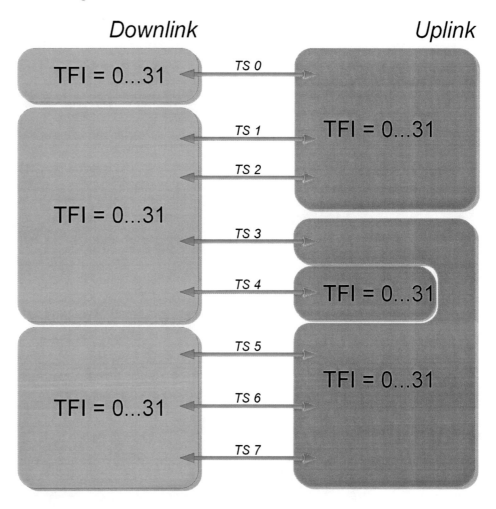

Downlink   *Uplink*

TFI = 0...31 — TS 0 — TFI = 0...31

TFI = 0...31 — TS 1 / TS 2 — TFI = 0...31

TFI = 0...31 — TS 3 — TFI = 0...31

TS 4 — TFI = 0...31

TFI = 0...31 — TS 5 / TS 6 / TS 7 — TFI = 0...31

The Trouble with Resource Allocation:

ö A packet switched mobile network does not deploy dedicated resources but rather, operates on a resource on demand principle. One PDCH or PDTCH is shared among many users.

ö The downlink direction is not critical since the network can identify the mobile station which is being addressed in each downlink block.

ö The uplink direction is extremely critical since transmissions from several mobile stations could collide.

Uplink Transmissions need to be scheduled and controlled by the network

GPRS\EGPRS Provides Three Uplink Resource Allocation Methods:

ö **Fixed Allocation of Uplink Resources**
a mandatory feature of both the network *) and the mobile station.

ö **Dynamic Allocation of Uplink Resources**
a mandatory feature of both the network *) and the mobile station.

ö **Extended Dynamic Allocation of Uplink Resources**
*an optional feature for the network and for most mobile stations and
a mandatory feature for mobile stations with multislot classes 22, 24, 25 and 27.*

*) Note that the network needs to support either fixed or dynamic resource allocation.

(1) The Fixed Allocation of Uplink Resources:

ö   The network conveys a bitmap to the mobile station. This identifies all blocks within several consecutive 52-multiframes in which the mobile station can transmit.

ö   Depending on the number of time slots that are in use, more than one bitmap may be conveyed.

ö   In addition to the allocation bitmap, the mobile station needs to know when to start transmitting ($\Leftrightarrow$ Starting Time).

ö   Whilst being involved in a fixed allocation, the mobile station can request more resources, if required.

(2) The Fixed Allocation of Uplink Resources:

Fixed Allocation of Uplink Resources

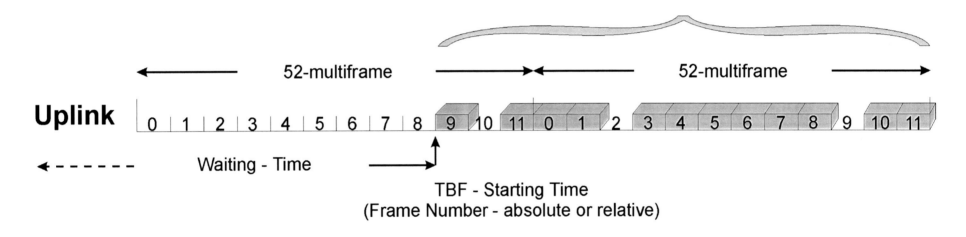

138

(1) The Dynamic Allocation of Uplink Resources:

- *The Dynamic Resource Allocation method is based on the use of the Uplink State Flag (USF).*

- *The USF is part of the MAC header of each bit of downlink data or of the control block that is sent.*

- *The USF of the downlink block k identifies the user of the uplink block (k+1)*

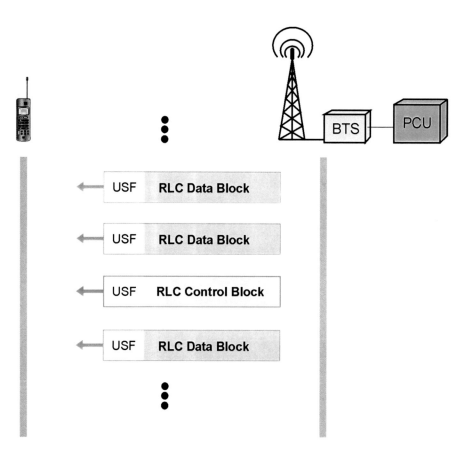

(2) The Dynamic Allocation of Uplink Resources:

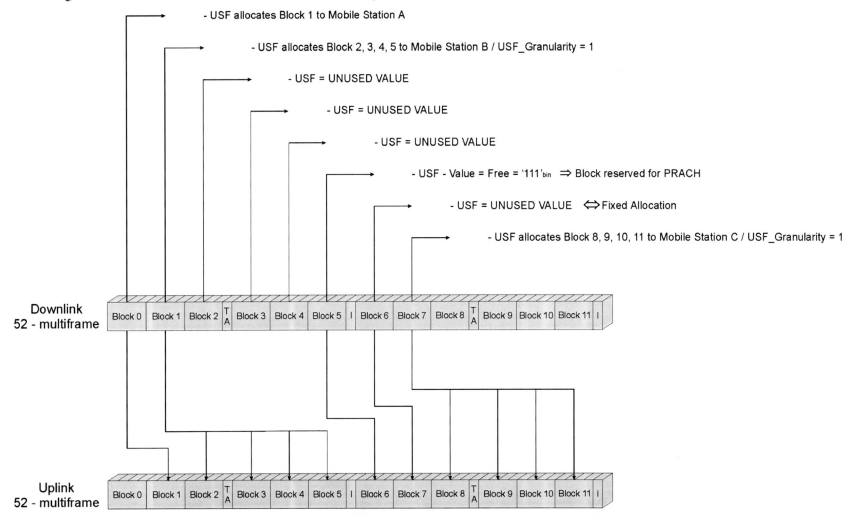

- USF allocates Block 1 to Mobile Station A

- USF allocates Block 2, 3, 4, 5 to Mobile Station B / USF_Granularity = 1

- USF = UNUSED VALUE

- USF = UNUSED VALUE

- USF = UNUSED VALUE

- USF - Value = Free = '111'$_{bin}$ ⇒ Block reserved for PRACH

- USF = UNUSED VALUE ⇔ Fixed Allocation

- USF allocates Block 8, 9, 10, 11 to Mobile Station C / USF_Granularity = 1

Downlink
52 - multiframe

| Block 0 | Block 1 | Block 2 | T A | Block 3 | Block 4 | Block 5 | I | Block 6 | Block 7 | Block 8 | T A | Block 9 | Block 10 | Block 11 | I |

Uplink
52 - multiframe

| Block 0 | Block 1 | Block 2 | T A | Block 3 | Block 4 | Block 5 | I | Block 6 | Block 7 | Block 8 | T A | Block 9 | Block 10 | Block 11 | I |

Part 3: Details of EGPRS
EDGE from A -Z Version 2.0

(1) More Details of the Uplink State Flag (USF):

ö **The USF has a fixed length of 3 bits**
Eight different values are possible.

ö **Eight, seven or six mobile stations can be distinguished**
depending on whether the PCCCH is part of that PDCH. Another USF-value is reserved to identify fixed allocations and for the USF-Granularity Flag = 0/1 (see previous slide).

ö **A mobile station, which is involved in a multislot assignment, most probably has various USFs assigned to it, one for each time slot.**

ö **Pay attention to the USF-GRANULARITY**
*parameter in the **PACK_UL_ASS** message. Depending on its value, a mobile station can not only use the next uplink block but the next four uplink blocks.*

(2) More Details about the Uplink State Flag (USF):

ö The USF is also used to allocate radio blocks for PRACHs.

The fixed value '111'$_{bin}$ is reserved to denote the PRACH.

ö Almost all active mobile stations have to decode all downlink data blocks in order to find "their" USF.

This imposes serious constraints on the power control method.

ö Mobile stations that intend to transmit on PRACH need to listen to the USF on that PDCH.

ö Mobile stations which are involved in a fixed allocation are not required to decode the USFs.

(1) The Extended Dynamic Allocation of Uplink Resources:

ö   is an optional allocation method for the network and the mobile station

ö   also uses the USF for uplink transmission scheduling.

ö   is only applicable to multislot assignments.

ö   relieves the multislot mobile station from "listening" to all downlink blocks on all assigned time slots.

ö   is explained in more detail on the following page.

(2) The Extended Dynamic Allocation of Uplink Resources:

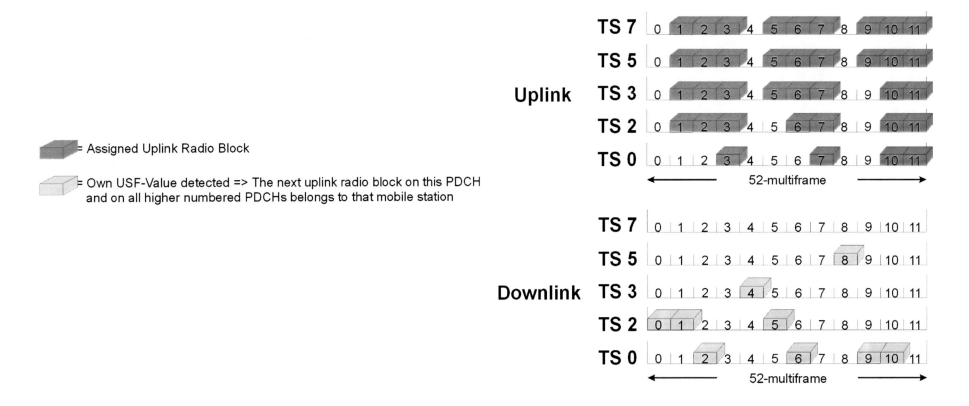

= Assigned Uplink Radio Block

= Own USF-Value detected => The next uplink radio block on this PDCH and on all higher numbered PDCHs belongs to that mobile station

Summary of Network Access and Resource Allocation:

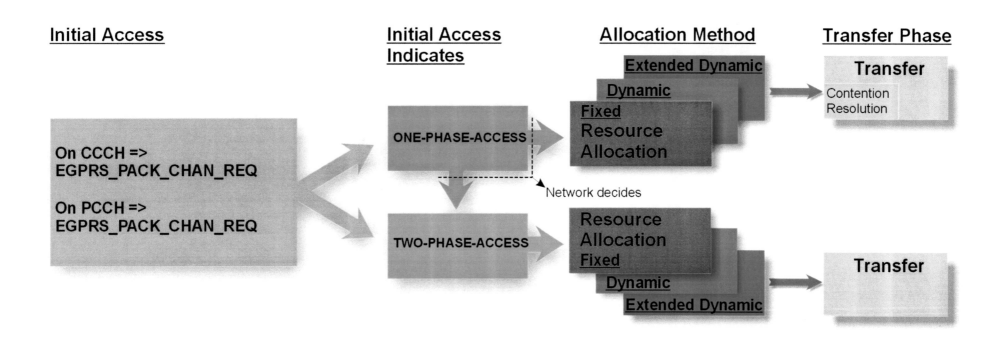

145

And what about Downlink Resource Allocation?

ö   Downlink transmission scheduling is not an issue since it is automatically controlled by the network.

ö   Downlink reception deserves some attention since a mobile station cannot determine in advance when packets will be sent by the network.

ö   *Thus, a mobile station needs to receive and decode all downlink data blocks on all assigned time slots whilst involved in a downlink TBF.*

ö   A mobile station will identify "its" downlink data packets by checking the TFI, which is part of each downlink block.

Operation of the Bi-directional PACCH:

ö   As opposed to PDTCHs, the relevant PACCH is bi-directional.

ö   Resources for issuing an RLC/MAC control message on PACCH need to be provided in a dynamic manner and on the network's demand (⇔ PCU).

ö   PACCH operation is different for the uplink and downlink directions.

and:

GMM and SM messages are *not* transmitted via the PACCH but via the PDTCH

(1) PACCH Operation for Downlink TBF's:

ö   Whilst involved in a downlink TBF, the mobile station needs to decode all data and control blocks on all assigned time slots.

Control messages on PACCH that are destined for a specific mobile station are tagged by:
- *including the downlink TFI in the RLC header of the RLC/MAC control message or by*
- *including the mobile station's identity (TLLI) or the downlink TFI within the message content.*

ö   At certain times, the mobile station needs to issue control messages on PACCH.

Note: a downlink TBF does not provide uplink resources to the mobile station. Accordingly, the network must allocate a single uplink block to a given mobile station dynamically. This procedure is explained on the next page.
Note: this type of operation is also required in the uplink direction when the release of an uplink TBF needs to be confirmed by the mobile station.

(2) PACCH Operation in Downlink Direction:

Network polls the mobile station

RRBP is the abbreviation for Relative Reserved Block Period.

The 2 bit long parameter is part of each RLC/MAC data and control block in downlink direction and allocates a single radio block in uplink direction to the addressed mobile station.

Dependent on the actual coding of RRBP, the mobile station needs to wait a defined number of TDMA frames before it is allowed to transmit its control messages on the same PDCH.

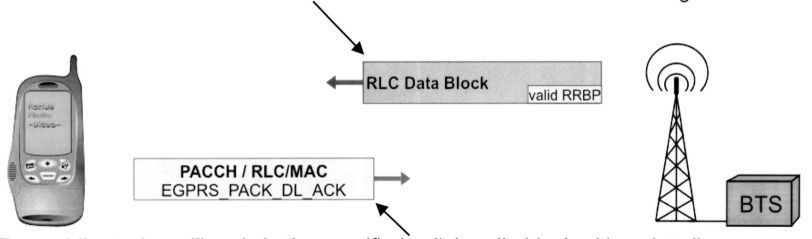

The mobile station will reply in the specified uplink radio block with a signaling message
(either EGPRS_PACK_DL_ACK or PACK_CTRL_ACK)

PACCH Operation for Uplink TBF's:

ö   Whilst involved in an uplink TBF, the mobile station may also use the allocated resources to transmit RLC/MAC control messages.

The distinction between RLC/MAC data and control blocks is made via the RLC/MAC header.

ö   The network can transmit RLC/MAC control messages to the mobile station on all assigned time slots at any time.

Thus, the mobile station must receive and decode all downlink blocks on all assigned time slots. The network will only limit the transmission of RLC/MAC control blocks on PACCH to the lowest numbered time slot (⇔ PDCH) for extended dynamic resource allocation. Distinguishing the RLC/MAC control messages for different mobile stations is carried out according to the procedure for downlink TBF PACCH operation.

Resource Release in GPRS\EGPRS:

ö   GPRS\EGPRS requires an active packet transaction (TBF), which can be terminated without any signaling effort.

ö   This requirement applies for both the uplink and downlink directions.

ö   Obviously, the release procedure for uplink and downlink TBFs is different.

151

(1) The Release of Uplink Resources:

ö   The release of the uplink resources is initiated by the mobile station.

ö   Uplink resource release is based on the countdown procedure.

ö   The countdown procedure is mandatory for all resource allocation methods, fixed, dynamic and extended dynamic.

ö   TBF release is confirmed by the network.

*Once the countdown procedure has ceased, the network will issue a **PACK_UL_ACK** with the Final Ack bit set to 1. This applies to both RLC operation modes, acknowledged and unacknowledged. Evidently, re-transmission can only be invoked in the acknowledged RLC operation mode.*

(2) The Release of Uplink Resources:

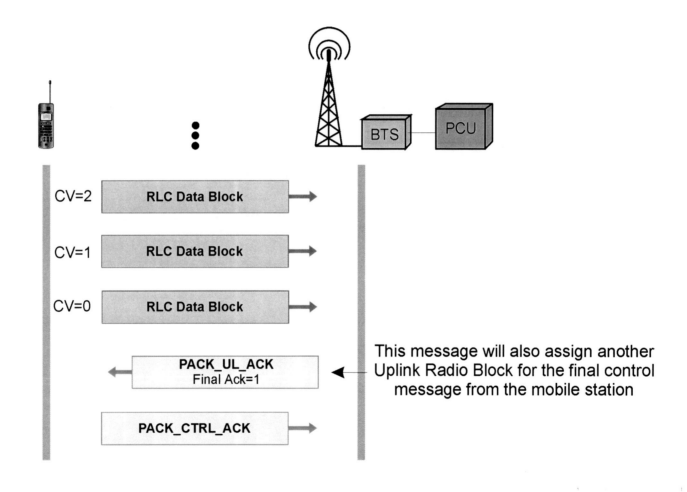

CV=2	RLC Data Block
CV=1	RLC Data Block
CV=0	RLC Data Block

PACK_UL_ACK
Final Ack=1

This message will also assign another Uplink Radio Block for the final control message from the mobile station

PACK_CTRL_ACK

(1) The Countdown Procedure:

ö   Every RLC/MAC uplink data block contains the 4 bit long parameter Countdown Value (CV).

ö   CV indicates the number of RLC/MAC data blocks that are still in the mobile station and need to be transmitted to the network.

ö   Before the countdown procedure starts, the value of CV defaults to '15'dec

ö   The countdown procedure can be started at '15'dec or at a smaller value.

*This depends on another parameter, BS_CV_MAX which is broadcast in **PACK_SYS_INFO 1**, **13** and **SYS_INFO 13**.*

(2) The Countdown Procedure:

ö   Once the countdown procedure has begun, the mobile station cannot invoke additional resources for that TBF.

This limitation does not apply if re-transmissions are necessary. In this case, the network must assign additional resources.

ö   The countdown procedure is somewhat different for single and multiple time slot assignments.

ö   The countdown procedure is based on the formula on the following page.

155

(3) The Countdown Procedure:

Countdown Procedure starts ì

$$\frac{(\text{Total No of Blocks - 1}) - \text{BSN}}{\text{No of Timeslots}} = 15 \text{ or BS\_CV\_MAX}$$

Note:
1. *The division is non-integer and its result needs to be rounded upwards.*
2. *This formula also applies in case of only one time slot*

BSN (Block Sequence Number):

- BSN numbers and identifies each RLC/MAC data block.

- BSN is used in both the acknowledged and unacknowledged RLC operation modes.

- BSN is a part of each RLC header in the uplink *and* downlink data blocks.

The Countdown Procedure in a One time slot Configuration:

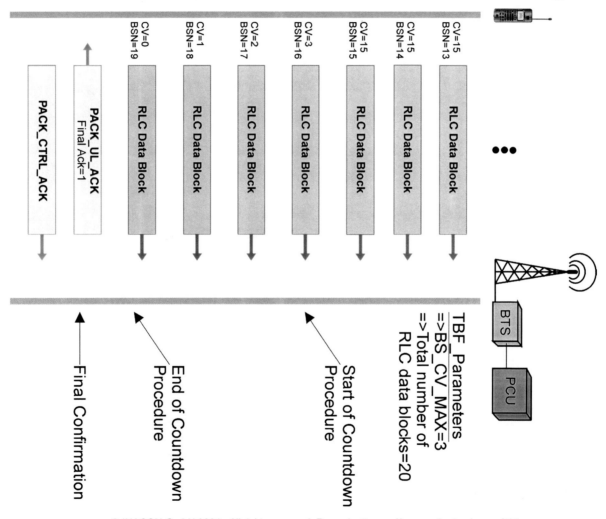

The Countdown Procedure in a 4 time slot Configuration:

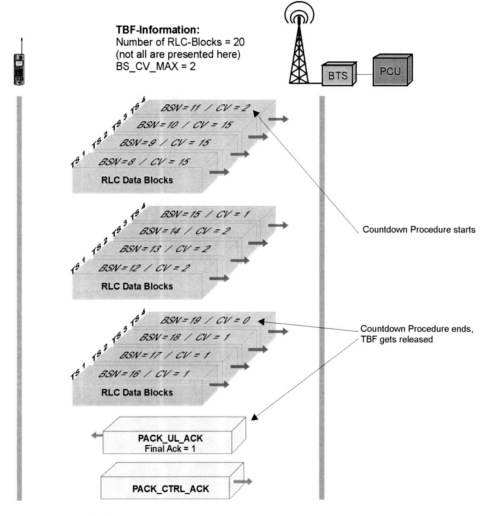

TBF-Information:
Number of RLC-Blocks = 20
(not all are presented here)
BS_CV_MAX = 2

BTS   PCU

TS 1 TS 2 TS 3 TS 4

BSN = 11 / CV = 2
BSN = 10 / CV = 15
BSN = 9 / CV = 15
BSN = 8 / CV = 15
RLC Data Blocks

BSN = 15 / CV = 1
BSN = 14 / CV = 2
BSN = 13 / CV = 2
BSN = 12 / CV = 2
RLC Data Blocks

BSN = 19 / CV = 0
BSN = 18 / CV = 1
BSN = 17 / CV = 1
BSN = 16 / CV = 1
RLC Data Blocks

Countdown Procedure starts

Countdown Procedure ends,
TBF gets released

PACK_UL_ACK
Final Ack = 1

PACK_CTRL_ACK

158

(1) The Release of Downlink Resources:

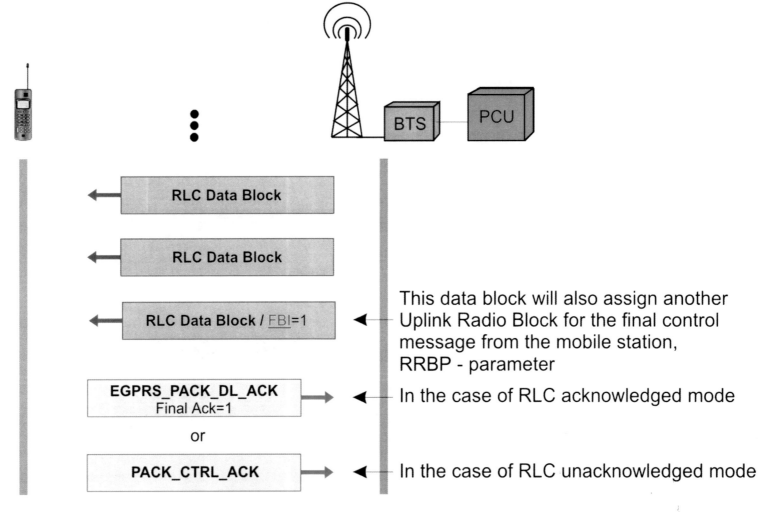

This data block will also assign another Uplink Radio Block for the final control message from the mobile station, RRBP - parameter

In the case of RLC acknowledged mode

In the case of RLC unacknowledged mode

(2) The Release of Downlink Resources:

ö   For downlink TBF release, the network sets the Final Block Indicator (FBI) bit to 1, which indicates that this block is the last one to be sent.

ö   The FBI bit is part of the RLC header in each downlink data block.

ö   Within the same block, the network will allocate a single uplink block to the mobile station to confirm whether proper reception has been made or whether re-transmission is required (⇔ acknowledged mode only).

ö   In the unacknowledged RLC mode, the mobile station will confirm the TBF release by issuing a **PACK_CTRL_ACK**.

Acknowledged ⇔ Unacknowledged Operation Mode:

As is the case for other GPRS\EGPRS protocols, the
Radio Link Control protocol (RLC) on the air interface
has two operation modes, namely:

ö   **Unacknowledged Operation Mode**

ö   **Acknowledged Operation Mode**

Details of Unacknowledged Operation Mode:

- RLC data block transfer in the RLC unacknowledged mode does not include re-transmission, except during the release of an uplink TBF in which the last transmitted uplink block may be retransmitted.

- The block sequence number (BSN) in the RLC data block header is used to number the RLC data blocks for reassembly. The receiving side sends Packet Ack/Nack messages in order to convey any other control signaling that is necessary (e.g. monitoring channel quality for downlink transfer or timing advance correction for uplink transfers).

- During RLC unacknowledged mode operation, LLC PDUs which have been received are delivered to the higher layer in the order in which they have been received. RLC data units which are not received are substituted by fill in bits with a value of '0'. However, for erroneous RLC data blocks for which the header is received correctly, the output from the decoder is delivered to the higher layer (ì LLC).

Details of the Acknowledged Operation Mode:

- In the EGPRS TBF mode, the transfer of RLC Data Blocks within the acknowledged RLC/MAC mode can be controlled by a selective **type I ARQ** mechanism, or by a **type II hybrid ARQ** (Incremental Redundancy: IR) mechanism, coupled with the numbering of the RLC Data Blocks within one Temporary Block Flow.

- The transfer of RLC data blocks in the RLC acknowledged mode uses the re-transmission of RLC data blocks. The transmitting side numbers the RLC data blocks via the block sequence number (BSN). The BSN is used for re-transmission and for reassembly. The receiving side sends PACKET Ack/Nack messages as a request for the re-transmission of RLC data blocks

- During RLC acknowledged mode operation, received LLC PDUs are delivered to the higher layer in the order in which they were originally transmitted.

Acknowledged RLC Operation Mode is based on the ARQ (**A**utomatic **R**epeat re**Q**uest) mechanism.

(1) Automatic Repeat Request (ARQ):

ö   Erroneous data blocks can be selected for re-transmission.

ö   Each frame is numbered in a fashion which is not ambiguous within a sliding receive / transmit window.

ö   The respective acknowledgement messages therefore contain a bitmap in order to properly distinguish received data blocks from erroneous data blocks.

(PACK_UL_ACK and *(EGPRS)_PACK_DL_ACK)*

(2) Automatic Repeat Request (ARQ):

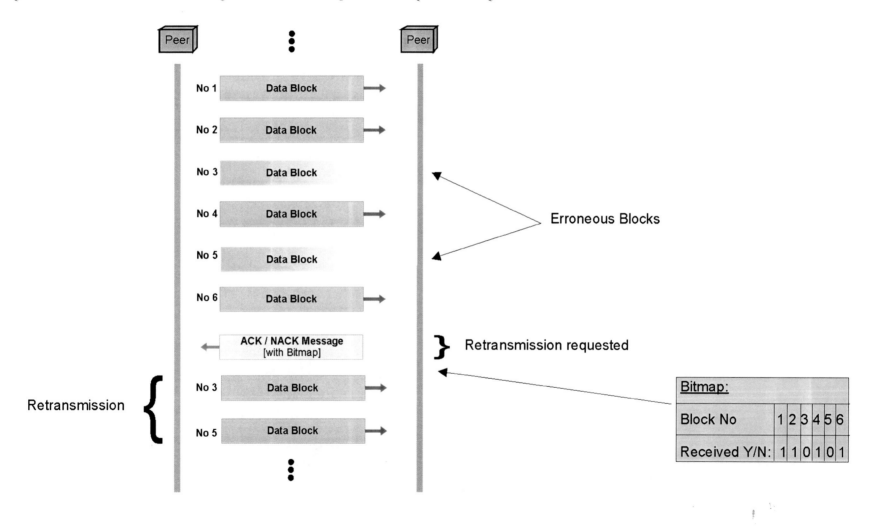

(1) Acknowledged Mode for RLC / MAC Operation:

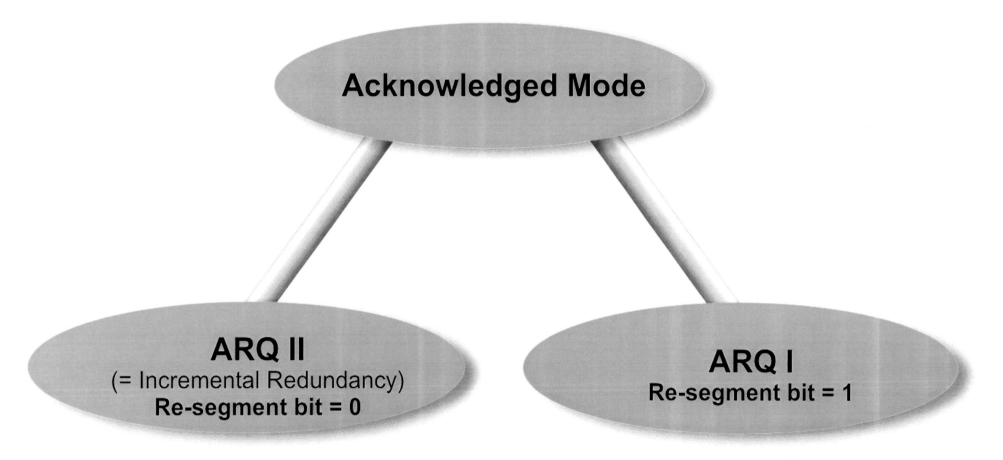

Note: ARQ II (Incremental Redundancy) is mandatory for the MS but optional for the network.

(2) Acknowledged Mode for RLC / MAC Operation:

ö The transfer of RLC Data Blocks in the acknowledged RLC/MAC mode can be controlled by a selective ARQ I (Automatic Repeat reQuest) mechanism or by an ARQ II(Incremental Redundancy: IR) mechanism.

ö <u>The re-segment bit within the Packet_UL_Ack or the Packet_UL_ Assignment message is used to set the ARQ modus in acknowledged mode to either ARQ I or ARQ II, Incremental Redundancy.</u>

ö The sending side (MS or Network) transmits blocks within a window (ì *Window Size*) and the receiving (MS or Network) site sends Packet Uplink Ack / Nack or Packet Downlink Ack / Nack messages when needed.

ö Depending on the quality of the link, an initial MCS is selected for an RLC block. For <u>re-transmission</u>, either the same or another MCS from the same family of MCSs can be selected (ì *Data Block Families*). For example, if MCS-7 is selected for the initial transmission of an RLC block, any MCS belonging to family B can be used for re-transmission.

The selection of the MCS is controlled by the network.

(3) Acknowledged Mode for RLC / MAC Operation:

ö **Window Size (WS):**

In EGPRS, the window size (WS) is set by the network according to the number of time slots allocated in the direction of the TBF (uplink or downlink).

The smallest WS is 64 Blocks. The WS can be set <u>in steps of 32 Blocks</u> up to the maximum (depending on the number of assigned time slots). Preferably, the selected window should have the maximum size.

Window size	Allocated timeslots (Multislot capability)							
Blocks	1	2	3	4	5	6	7	8
minimum: 64								
192	Max	permitted window size						
256		Max						
384			Max					
512				Max				
640					Max			
768						Max		
896							Max	
1024								Max

(4) Acknowledged Mode for RLC / MAC Operation:

ö The window size can be set independently on uplink and downlink. The MS supports the maximum window size that corresponds to the respective multi-slot capability. The selected WS is indicated within PACKET UL/DL ASSIGNMENT and PACKET time slot RECONFIGURE.

ö Once a window size has been selected for a given MS, it may be increased in size but not reduced in order to prevent dropping data blocks from the window.

ö Note: If a TBF is reallocated such that the number of allocated time slots is reduced, the window size can become larger than the maximum WS for the new resources e.g. If three TSs are allocated, the WS is 384 blocks and if the number of allocated TS is reduced to two, the WS will still remain at 384 blocks.

(1) Incremental Redundancy:

ö The Incremental Redundancy mode of acknowledged RLC operation in EGPRS has three major advantages when compared to GPRS RLC acknowledged mode:

1.) In GPRS, once a data block has been transmitted with a designated coding scheme, it can only be retransmitted with the same coding scheme. This also applies when the network requests a coding scheme with higher protection against transmission errors.

Every MCS out of the same data block family can be used for re-transmission in EGPRS.

Incremental Redundancy reduces re-transmission on the air interface and therefore increases the data throughput.

(2) Incremental Redundancy:

2.) In GPRS, decoding a retransmitted data block is based solely on this data block alone. All previously received versions of the data block are not mentioned.

In the EGPRS Incremental Redundancy mode, the versions of the data block which were received initially and all retransmitted data blocks are decoded jointly.

Incremental Redundancy also decodes the information when the retransmitted RLC data Block includes transmission errors.

(3) Incremental Redundancy:

3.) In GPRS, re-transmission does not provide the receiver with any new information (same coding and puncturing scheme as before).

In EGPRS, a different coding scheme is used for the re-transmission of each of a data block. Hence, re-transmission increases the number of error free bits.

When using the Incremental Redundancy mode with each re-transmission, the number of error free bits of a received data block increases.

(4) Incremental Redundancy:

ö The MCS selection may take into account the Incremental Redundancy capability of the receiver by selecting an MCS with higher data throughput but less error correction capability.

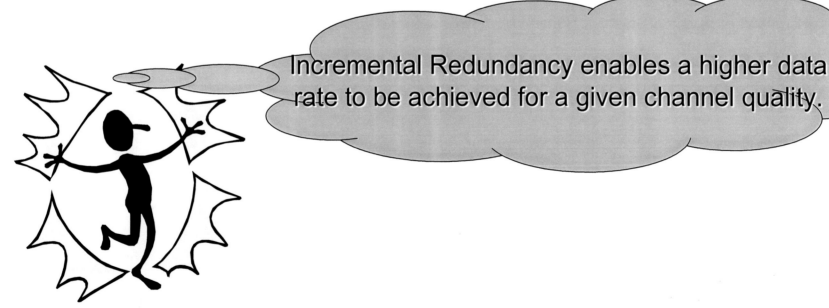

Incremental Redundancy enables a higher data rate to be achieved for a given channel quality.

173

(5) Incremental Redundancy:

When re-transmission takes place, the MCS is determined by both the MCS which were used initially and the newly requested MCS for future data block transmission, according to the table below.

e.g: MCS-8 for initial transmission
MCS-4 requested for subsequent data blocks
MCS for re-transmission = MCS-6

re-segment bit=0 (ARQ II Incremenal Redundancy)

MCS used for initial Transmission	Commanded MCS after Initial Transmission								
	MCS-9	MCS-8	MCS-7	MCS-6	MCS-5	MCS-4	MCS-3	MCS-2	MCS-1
MCS-9	MCS-9	MCS-6	MCS-6	MCS-6	MCS-6	MCS-6	MCS-6	MCS-6	MCS-6
MCS-8	MCS-8	MCS-8	MCS-6	MCS-6	MCS-6	MCS-6	MCS-6	MCS-6	MCS-6
MCS-7	MCS-7	MCS-7	MCS-7	MCS-5	MCS-5	MCS-5	MCS-5	MCS-5	MCS-5
MCS-6	MCS-9	MCS-6	MCS-6	MCS-6	MCS-6	MCS-6	MCS-6	MCS-6	MCS-6
MCS-5	MCS-7	MCS-7	MCS-7	MCS-5	MCS-5	MCS-5	MCS-5	MCS-5	MCS-5
MCS-4	MCS-4								
MCS-3	MCS-3								
MCS-2	MCS-2								
MCS-1	MCS-1								

(6) Incremental Redundancy:

ö For Incremental Redundancy, the Puncturing Scheme changes for re-transmission.

ö If MCS for re-transmission differs from the initial one, a PS will be used. This guarantees a minimum cross correlation.

ö Since different bit positions are deleted in different Puncturing Schemes, the number of known bits at the receiver increases with each re-transmission.

(7) Incremental Redundancy:

MCS for initial transmission	MCS for retransmission	PS used for initial transmission	PS used for retransmission
MCS-9	MCS-6	PS 1 or PS 3	PS 1
		PS 2	PS 2
MCS-6	MCS-9	PS 1	PS 3
		PS 2	PS 2
MCS-7	MCS-5	any	PS 1
MCS-5	MCS-7	any	PS 2
all other combinations		any	PS 1

Note: The PS that must be used can also be the same as the previous one. However, minimum cross correlation is still guaranteed due to puncturing at different positions.

(8) Incremental Redundancy:

re-segment bit =0

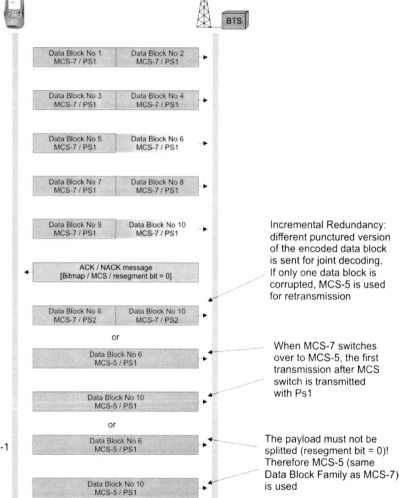

Commanded MCS:

MCS-9 / MCS-8 / MCS-7

or

MCS-6 / MCS-5

or

MCS-4 / MCS-3 / MCS-2 / MCS-1

Data Block No 1
MCS-7 / PS1

Data Block No 2
MCS-7 / PS1

Data Block No 3
MCS-7 / PS1

Data Block No 4
MCS-7 / PS1

Data Block No 5
MCS-7 / PS1

Data Block No 6
MCS-7 / PS1

Data Block No 7
MCS-7 / PS1

Data Block No 8
MCS-7 / PS1

Data Block No 9
MCS-7 / PS1

Data Block No 10
MCS-7 / PS1

ACK / NACK message
[Bitmap / MCS / resegment bit = 0]

Data Block No 6
MCS-7 / PS2

Data Block No 10
MCS-7 / PS2

or

Data Block No 6
MCS-5 / PS1

Data Block No 10
MCS-5 / PS1

or

Data Block No 6
MCS-5 / PS1

Data Block No 10
MCS-5 / PS1

Incremental Redundancy:
different punctured version
of the encoded data block
is sent for joint decoding.
If only one data block is
corrupted, MCS-5 is used
for retransmission

When MCS-7 switches
over to MCS-5, the first
transmission after MCS
switch is transmitted
with Ps1

The payload must not be
splitted (resegment bit = 0)!
Therefore MCS-5 (same
Data Block Family as MCS-7)
is used

(1) ARQ I:

ö   In the EGPRS ARQ I, the operation is similar to that of EGPRS ARQ II, the difference being that there is no incremental redundancy.

ö   In ARQ 1, error correction is only done by block re-transmission; erroneous blocks are not stored.

ö   However, since no joint decoding is carried out, a retransmitted block can also be split into two blocks according to requested MCS within the same data block family.

ö   All the other procedures e.g. different PS´s for initial transmission and re-transmission are exactly the same as those for Incremental Redundancy.

(2) ARQ I:

E.g.: MCS-6 for initial transmission
MCS-1 requested for subsequent data blocks
MCS for re-transmission = MCS-3
⇨ **RLC data block is split.**

re-segment bit=1 (ARQ I)

MCS used for initial Transmission	Commanded MCS after Initial Transmission								
	MCS-9	MCS-8	MCS-7	MCS-6	MCS-5	MCS-4	MCS-3	MCS-2	MCS-1
MCS-9	MCS-9	MCS-6	MCS-6	MCS-6	MCS-3	MCS-3	MCS-3	MCS-3	MCS-3
MCS-8	MCS-8	MCS-8	MCS-6	MCS-6	MCS-3	MCS-3	MCS-3	MCS-3	MCS-3
MCS-7	MCS-7	MCS-7	MCS-7	MCS-5	MCS-5	MCS-2	MCS-2	MCS-2	MCS-2
MCS-6	MCS-9	MCS-6	MCS-6	MCS-6	MCS-3	MCS-3	MCS-3	MCS-3	MCS-3
MCS-5	MCS-7	MCS-7	MCS-7	MCS-5	MCS-5	MCS-2	MCS-2	MCS-2	MCS-2
MCS-4	MCS-4	MCS-4	MCS-4	MCS-4	MCS-4	MCS-4	MCS-1	MCS-1	MCS-1
MCS-3	MCS-3								
MCS-2	MCS-2								
MCS-1	MCS-1								

(3) ARQ I:

re-segment bit =1

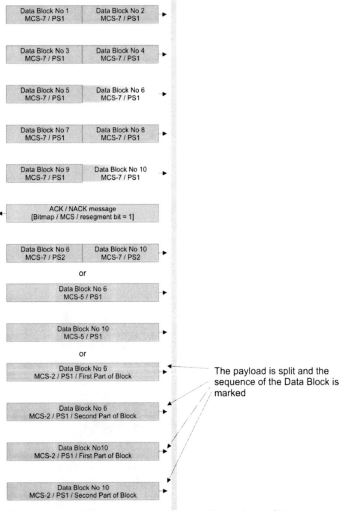

Commanded MCS:

MCS-9 / MCS-8 / MCS-7

or

MCS-6 / MCS-5

or

MCS-4 / MCS-3/ MCS-2 / MCS-1

The payload is split and the sequence of the Data Block is marked

(1) Timing Advance Control in EGPRS using PTCCH
(Continuous Timing Advance Update Procedure)

ö The PTCCH/U is divided into 16 subchannels within eight 52-multiframes.

ö The 16 subchannels can be assigned to 16 different active mobile stations.

ö Obviously, each PTCCH/U subchannel is repeated every 416 TDMA frames (1.92 s)

ö The active mobile stations transmit one access burst with TA = 0 to the BTS once every eight 52-multiframes within "their" subchannel.

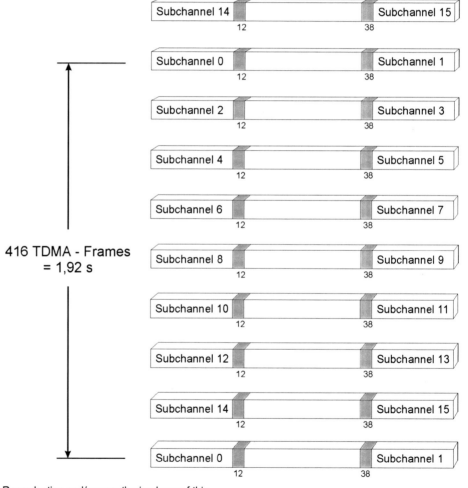

416 TDMA - Frames = 1,92 s

(2) Timing Advance Control in EGPRS using PTCCH (Continuous Timing Advance Update Procedure)

ö Based on the PTCCH/U transmission, the network can recalculate the timing advance value.

ö The updated TA values can be conveyed to a maximum of 16 different mobile stations in the TA message.

ö The TA message is carried in 4 consecutive PTCCH/D frames as illustrated in the figure on the following page. The format of the TA message is shown below.

	Octet 1		Octet 2		Octet 16	Octet 17		Octet 23
0	TA - Value for TAI = 0	0	TA - Value for TAI = 1	••••	0	TA - Value for TAI = 15	00101011	••••	00101011

| 1 | ←——7 bit——→ | 1 | ←——7 bit——→ | | 1 | ←——7 bit——→ | ←——— 7 Fill Octets ———→ |

←——————————————— 184 bit ———————————————→

(3) Timing Advance Control in EGPRS using PTCCH
(Continuous Timing Advance Update Procedure)

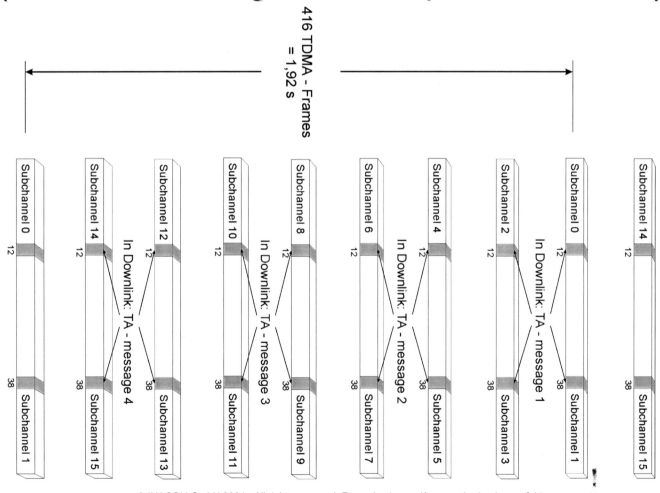

Other Options:

As illustrated, the network can also use RLC/MAC control messages on PACCH to convey updated TA values to a mobile station.

In this case, the mobile station needs to send shortened PACK_CTRL_ACK's to the network when polled (4 identical access bursts) in the specified uplink radio block.

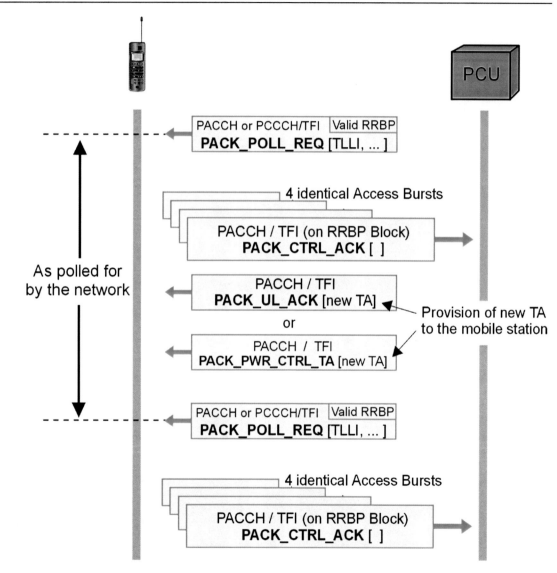

184

QoS in EGPRS

185

Quality of Service (QoS):

ö   In EGPRS, every subscription is related to a QoS profile.

ö   Before packet data transmission can be carried out, the network (GGSN) and the mobile station must negotiate a QoS profile.

EGPRS provides two options:

ö   The network operator can define a default QoS profile that is applicable to all subscribers (⇔ Best Effort)

ö   The network operator can define several QoS profiles that can be subscribed to.

The QoS Profile:

The QoS profile is considered to be a single parameter with multiple data transfer parameters.

The four following QoS Classes are defined in Release 99:

z   Conversational Class

z   Streaming Class

z   Interactive Class

z   Background Class

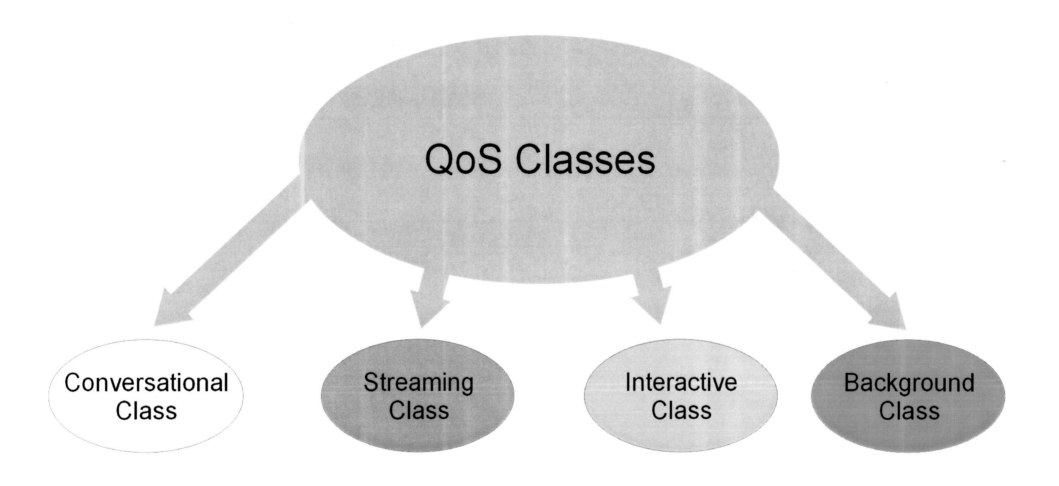

188

The Conversational Class

z   This QoS Class is used to carry real time applications such as video telephony, voice over IP and video conferencing which are very sensitive to delay

z   Real time conversation is always carried out between peers, or groups, of live (human) consumers.

z   In this QoS Class, the characteristics required are determined by human perception.

The fundamental characteristics of the Conversational Class are:

preserve time relation (variation) between information entities of the stream,

stringent and low delay.

The Streaming Class

z   This QoS Class is used to carry applications such as real time video and audio which are highly sensitive to delay

z   The real time data flow is always aimed at a living human destination. *It is one way transport.*

z   Since the stream is usually aligned to time at the receiving end, the largest acceptable delay variation over the transmission media is determined by the capacity of the time alignment function of the application. Acceptable delay variation is thus much greater than the delay variation determined by the limits of human perception (ø Conversational Class).

The fundamental characteristics of the Streaming Class are:

preserve time relation, variation, between information entities of the stream,

although it does not have any requirements on low transfer delay.

190

The Interactive Class

z   This QoS Class is used for applications such as browsing the web, data base retrieval, server access etc.

z   At its destination, the message response is expected within a certain time.

z   The content of the packets are be transferred transparently with a low bit error rate.

The fundamental characteristics of the Conversational Class are:

limited request - response delay.

preserve payload content.

The Background Class

z   This QoS Class is used for applications such as E-mail, SMS, downloading from databases, etc.

z   At its destination, the message, response, is not expected within a certain time. Therefore, this scheme is fairly insensitive to time.

z   The content of the packets are transferred transparently with a low bit error rate.

The fundamental characteristics of the Background Class are:

that the destination does not expect to receive the data within a certain period of time.

preserve payload content.

(1) Description of Release 99 QoS Attributes:

For the description of the QoS Classes in Release 99, the following attributes are defined:

- z Traffic Class.
- z Maximum bit rate.
- z Guaranteed bit rate.
- z Delivery order.
- z Maximum SDU size.
- z SDU format information.
- z SDU error ratio.
- z Residual bit error ratio.
- z Delivery of erroneous SDU.
- z Transfer delay.
- z Traffic handling priority.
- z Allocation/Retention Priority.

193

(2) Description of Release 99 QoS Attributes:

z **Traffic Class**

The Traffic Class is no different to the QoS profile which is negotiated.
By including the traffic class itself as an attribute, the Network can make
assumptions about the traffic source and optimise the transport for that type
of traffic.

z **Maximum bit rate (kbps)**

This defines the maximum number of bits delivered from Service Access
Point (SAP) to SAP within a certain period of time divided by its duration.

(3) Description of Release 99 QoS Attributes:

z **Guaranteed bit rate (kbps)**

This determines the guaranteed number of bits delivered from SAP to SAP within a period of time divided by its duration.

z **Delivery order (y/n)**

This indicates whether the Bearer service provides in sequence SDU delivery or not.

z **Maximum SDU size (octets)**

This determines the maximum permitted size of SDUs.
This attribute is used for admission control and policing.

195

(4) Description of Release 99 QoS Attributes:

z **SDU format information (bits)**

This informs the network about the exact size of the SDU.

This value is needed in order to adapt the RLC PDU to the bearer SDU size for the RLC unacknowledged mode .

The exact syntax of SDU format information attribute has to be defined.

z **SDU error ratio**

This attribute indicates the fraction of SDUs which are lost or detected as erroneous.

The SDU error ratio is only set for data flows for which error detection has been requested.

For the Interactive and Background classes, the SDU error ratio is used as a target value.

(5) Description of Release 99 QoS Attributes:

z **Residual bit error ratio**

This indicates the undetected bit error ratio for the data flow.

z **Delivery of erroneous SDUs (y/n/-)**

This indicates whether SDUs with detected errors will be delivered or not.

z **Transfer delay (ms)**

The delay of a data packet is defined as the time which elapses between its request for transfer and its delivery at the other SAP.

197

(6) Description of Release 99 QoS Attributes:

z **Traffic handling priority**

Only used for the Interactive Class.

This attribute specifies the relative importance of handling the data flow in comparison to other data flows that use bearers within the Interactive Class.

z **Allocation/Retention Priority**

This attribute specifies the relative importance of the data flow for the allocation and retention of the Radio access bearer. The Allocation/Retention Priority attribute is a subscription parameter which is not negotiated by the mobile terminal.

(1) Overview of Release 99 QoS Attributes:

The following table summarises the defined QoS attributes and their relevance to each traffic class.

Note that the traffic class is an attribute in itself.

Traffic class	Conversational class	Streaming class	Interactive class	Background class
Maximum bitrate	X	X	X	X
Delivery order	X	X	X	X
Maximum SDU size	X	X	X	X
SDU format information	X	X		
SDU error ratio	X	X	X	X
Residual bit error ratio	X	X	X	X
Delivery of erroneous SDUs	X	X	X	X
Transfer delay	X	X		
Guaranteed bit rate	X	X		
Traffic handling priority			X	
Allocation/Retention priority	X	X	X	X

(2) Overview of Release 99 QoS Attributes:

ö The following table lists the value ranges of the Qos attributes.

ö There are many possible QoS profiles defined by the combinations of the attributes.

ö A network can only support a limited subset of possible QoS profiles.

ö An MS can request a value for each of the QoS attributes. The network negotiates each attribute to a level that it can handle.

Note that the discussion on QoS attributes is continuing and not all attributes are clearly defined.

(3) Overview of Release 99 QoS Attributes:

Traffic class	Conversational class	Streaming class	Interactive class	Background class
Maximum bitrate (kbps)	< 2 048	< 2 048	< 2 048	< 2 048
Delivery order	Yes/No	Yes/No	Yes/No	Yes/No
Maximum SDU size (octets)	<=1 500 or 1 502	<=1 500 or 1 502	<=1 500 or 1 502	<=1 500 or 1 502
SDU format information	(to be defined)	(to be defined)		
Delivery of erroneous SDUs	Yes/No/-	Yes/No/-	Yes/No/-	Yes/No/-
Residual BER	$5*10^{-2}, 10^{-2}, 5*10^{-3}, 10^{-3}, 10^{-4}, 10^{-6}$	$5*10^{-2}, 10^{-2}, 5*10^{-3}, 10^{-3}, 10^{-4}, 10^{-5}, 10^{-6}$	$4*10^{-3}, 10^{-5}, 6*10^{-8}$	$4*10^{-3}, 10^{-5}, 6*10^{-8}$
SDU error ratio	$10^{-2}, 7*10^{-3}, 10^{-3}, 10^{-4}, 10^{-5}$	$10^{-1}, 10^{-2}, 7*10^{-3}, 10^{-3}, 10^{-4}, 10^{-5}$	$10^{-3}, 10^{-4}, 10^{-6}$	$10^{-3}, 10^{-4}, 10^{-6}$
Transfer delay (ms)	80 – maximum value	250 – maximum value		
Guaranteed bit rate (kbps)	< 2 048	< 2 048		
Traffic handling priority			1,2,3	
Allocation/Retention priority	1,2,3	1,2,3	1,2,3	1,2,3

The Release 97/98 QoS Profile:

The R97/98 QoS profile is built up using the following parameters:

ö **Service Precedence / Priority.**
Does a subscriber enjoy transmission precedence?

ö **Delay Class.**
How much time will a data packet need in order to be routed through the EGPRS network?

ö **Throughput.**
Mean Throughput and Peak Throughput need to be distinguished.

ö **Reliability Class.**
How much error control and correction is provided by EGPRS?

Service Precedence / Priority:

- Under normal network conditions, all users will be served equally. However, in the case of network congestion, those users with a higher priority level will have their transactions treated with a higher priority than those with a lower priority.
- Thus, a user with a low priority level may encounter higher delay periods or even loss of data if network overload occurs.
- Three levels of precedence are defined: '1' having the highest priority and '3' having the lowest.

Precedence	Precedence Name	Interpretation
1	High priority	Service commitments shall be maintained ahead of precedence classes 2 and 3.
2	Normal priority	Service commitments shall be maintained ahead of precedence classes 3.
3	Low priority	Service commitments shall be maintained after precedence classes 1 and 2.

(1) Delay Class:

- As its name suggests, the delay class relates to the maximum delay times that a data packet may encounter whilst being transported through the EGPRS network.

- Note that this delay does not account for periods of delay that are caused by effects outside the PLMN.

- Four delay classes need to be distinguished with '1' offering the lowest delay periods and '3' bearing the highest risk of delay.

- Delay class '4' relates to 'best effort' which means that all transactions are handled according to the "first-in-first-out" principle.

(2) Delay Class:

Delay Class	Delay (maximum values)			
	SDU size: 128 octets		SDU size: 1024 octets	
	Mean Transfer Delay (sec)	95 percentile Delay (sec)	Mean Transfer Delay	95 percentile Delay (sec)
1. (Predictive)	< 0.5	< 1.5	< 2	< 7
2. (Predictive)	< 5	< 25	< 15	< 75
3. (Predictive)	< 50	< 250	< 75	< 375
4. (Best Effort)	Unspecified			

SDU = Service Data Unit

(1) Mean Throughput Rate:

- The mean throughput rate is measured in units of octets per hour and therefore represents an average value.

- No less than 19 different mean throughput rates have been defined, ranging from 0.22 bit/s up to 111 kbit/s.

- As for the delay class parameter, there is an extra mean throughput rate named 'best effort'.

(2) Mean Throughput Rate:

Mean Throughput Class	Mean Throughput in octets per hour
1	100 (~0.22 bit/s).
2	200 (~0.44 bit/s).
3	500 (~1.11 bit/s).
4	1 000 (~2.2 bit/s).
5	2 000 (~4.4 bit/s).
6	5 000 (~11.1 bit/s).
7	10 000 (~22 bit/s).
8	20 000 (~44 bit/s).
9	50 000 (~111 bit/s).
10	100 000 (~0.22 kbit/s).
11	200 000 (~0.44 kbit/s).
12	500 000 (~1.11 kbit/s).
13	1 000 000 (~2.2 kbit/s).
14	2 000 000 (~4.4 kbit/s).
15	5 000 000 (~11.1 kbit/s).
16	10 000 000 (~22 kbit/s).
17	20 000 000 (~44 kbit/s).
18	50 000 000 (~111 kbit/s).
31	Best effort

Peak Throughput Rate:

- The peak throughput rate is measured in units of octets per second.

- 9 different peak throughput rates are defined offering transfer rates from 8 kbit/s up to an impressive 2.048 Mbit/s.

PEAK Throughput Class	Peak Throughput (Octets / s)
1	Up to 1000 (8 kbits/s).
2	Up to 2000 (16 kbits/s).
3	Up to 4000 (32 kbits/s).
4	Up to 8000 (64 kbits/s).
5	Up to 16000 (128 kbits/s).
6	Up to 32000 (256 kbits/s).
7	Up to 64000 (512 kbits/s).
8	Up to 128000 (1024 kbits/s).
9	Up to 256000 (2048 kbits/s).

(1) Reliability Class:

- In this context, reliability relates to the probability of data loss, data corruption or out-of-sequence delivery of data packets.

- Five different reliability classes have been defined, whereas the differences among them are due to data protection measures that are applied by the underlying EGPRS protocols such as LLC or RLC/MAC.

- The table on the following page outlines the interdependence of the various reliability classes and the EGPRS protocols.

209

(2) Reliability Class:

Reliability Class	GTP Mode	LLC Frame Mode	LLC Data Protection	RLC Block Mode	Traffic Type
1	Acknowledged	Acknowledged	Protected	Acknowledged	Non real-time traffic, error-sensitive application that cannot cope with data loss.
2	Unacknowledged	Acknowledged	Protected	Acknowledged	Non real-time traffic, error-sensitive application that can cope with infrequent data loss.
3	Unacknowledged	Unacknowledged	Protected	Acknowledged	Non real-time traffic, error-sensitive application that can cope with data loss, GMM/SM, and SMS.
4	Unacknowledged	Unacknowledged	Protected	Unacknowledged	Real-time traffic, error-sensitive application that can cope with data loss.
5	Unacknowledged	Unacknowledged	Unprotected	Unacknowledged	Real-time traffic, error non-sensitive application that can cope with data loss.

(1) Interworking between R97/98 and R99

ö   For the purpose of interworking between different releases, mapping rules between GPRS Release 97/98 (R97/98) and GPRS Release 99 (R99) have been defined.

ö   The overall principle of mapping between two profiles is as follows.
Two profiles, each applied in their respective network release, provide the same or at least similar QoS. The GPRS R97/98 equipment cannot provide a real time service which corresponds to the R99 conversational and streaming traffic classes. Therefore, mapping is always directed to the non-real time interactive and background traffic classes.

(2) Interworking between R97/98 and R99

ö   Mapping is always performed by the SGSN.

ö   The Air interface Session Management and GTP messages of R99 contain the R99 attributes as an extension of the R97/98 QoS Information Element.

ö   Thus, no mapping is required if an MS R97/98 visits an SGSN R99 and the GGSN corresponds to release 97/98 or 99.

ö   Mapping is only required if an R99 MS is served by an R99 SGSN and the GGSN corresponds to release 97/98. In this case the *request QoS* profile is mapped by the SGSN into an R97/98 Qos profile and the *negotiated Qos* profile which is returned by the GGSN is mapped back to R99.

ö   The following tables illustrate that the determined R99 QoS attributes from R97/98 attributes and the determined R97/98 QoS attributes from R99 attributes.

(3) Interworking between R97/98 and R99

Resulting R99 Attribute		Derived from R97/98 Attribute	
Name	Value	Value	Name
Traffic class	Interactive	1, 2, 3	Delay class
	Background	4	
Traffic handling priority	1	1	Delay class
	2	2	
	3	3	
SDU error ratio	10^{-6}	1, 2	Reliability class
	10^{-4}	3	
	10^{-3}	4, 5	
Residual bit error ratio	10^{-5}	1, 2, 3, 4	Reliability class
	$4*10^{-3}$	5	
Delivery of erroneous SDUs	'no'	1, 2, 3, 4	Reliability class
	'yes'	5	
Maximum bitrate [kbps]	8	1	Peak throughput class
	16	2	
	32	3	
	64	4	
	128	5	
	256	6	
	512	7	
	1024	8	
	2048	9	
Allocation/Retention priority	1	1	Precedence class
	2	2	
	3	3	
Delivery order	yes'	yes'	Reordering Required (Information in the SGSN and the GGSN PDP Contexts)
	'no'	'no'	
Maximum SDU size	1 500 octets	(Fixed value)	

(4) Interworking between R97/98 and R99

Resulting R97/98 Attribute		Derived from R99 Attribute	
Name	Value	Value	Name
Delay class	1	conversational	Traffic class
	1	streaming	Traffic class
	1	Interactive	Traffic class
		1	Traffic handling priority
	2	Interactive	Traffic class
		2	Traffic handling priority
	3	Interactive	Traffic class
		3	Traffic handling priority
	4	Background	Traffic class
Reliability class	2	$<= 10^{-5}$	SDU error ratio
	3	$10^{-5} < x <= 5*10^{-4}$	SDU error ratio
	4	$> 5*10^{-4}$	SDU error ratio
		$<= 2*10^{-4}$	Residual bit error ratio
	5	$> 5*10^{-4}$	SDU error ratio
		$> 2*10^{-4}$	Residual bit error ratio
Peak throughput class	1	< 16	Maximum bitrate [kbps]
	2	16 <= x < 32	
	3	32 <= x < 64	
	4	64 <= x < 128	
	5	128 <= x < 25	
	6	256 <= x < 512	
	7	512 <= x < 1024	
	8	1024 <= x < 2048	
	9	>= 2048	
Precedence class	1	1	Allocation/retention priority
	2	2	
	3	3	
Mean throughput class	Always set to 31	-	
Reordering Required (Information in the SGSN and the GGSN PDP Contexts)	yes'	yes'	Delivery order
	'no'	'no'	

The

EGPRS Protocol Stack

Table of Contents

- Radio Link Control / Medium Access Control (RLC / MAC)

- Logical Link Control (LLC)

- SubNetwork Dependent Convergence Protocol (SNDCP)

- Frame Relay / Network Service and Base Station Subsystem (E)GPRS Protocol (BSSGP)

- GPRS Tunneling Protocol

- Mobility and Session Management in EGPRS

The EGPRS Protocol Stack:

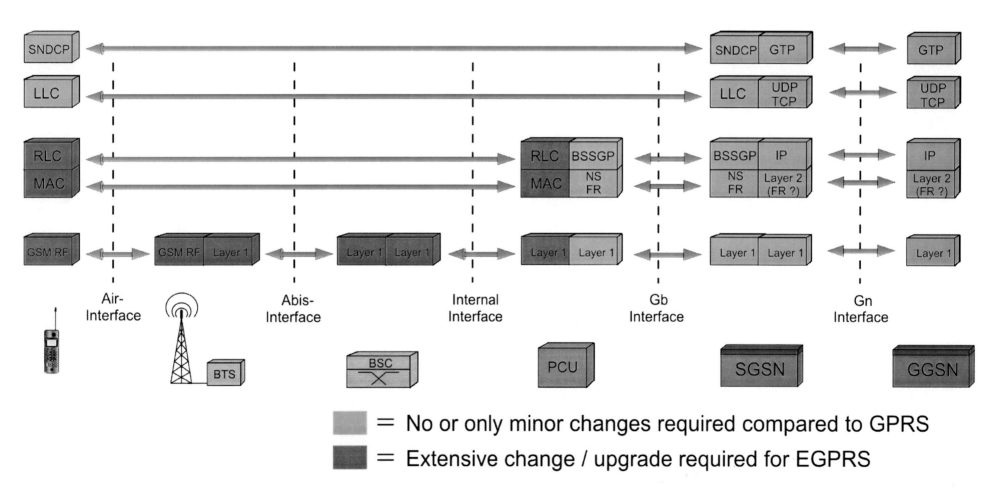

The Implications of the Previous Slide:

ö   Obviously, the BSC and the BTS are almost transparent to the EGPRS. The PCU carries out most of the packet related tasks concerning the BSS.
Note that the MSC is not included at all.

ö   The SGSN and the mobile station are the main peers within the EGPRS protocol stack.

ö   Note that the LLC is the lowest EGPRS protocol which is independent of the underlying air interface standard ($\Leftrightarrow$ IS-136).

(1) Radio Link Control / Medium Access Control (RLC / MAC):

ö **Tasks and Objectives of MAC:**

As *medium access control* suggests, the MAC sublayer is in charge of controlling the access of a device to a given transmission medium. Obviously, every protocol requires a more or less extensive MAC function. In the case of EGPRS, MAC is applied to the air interface where it deals with tasks such as access, sharing and release of the physical medium. When set against the OSI reference model, the tasks carried out by the MAC sublayer best match OSI layer 2.

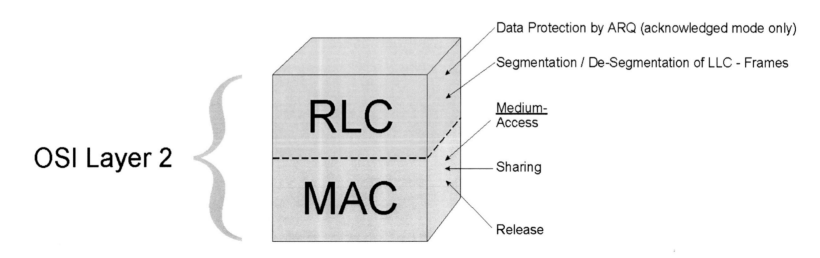

(2) Radio Link Control / Medium Access Control (RLC / MAC):

ö **Tasks and Objectives of RLC:**

Radio link control deals with the tasks of segmentation and de-segmentation of data units from higher layers (ì LLC). In the acknowledged operation mode, RLC also ensures that the data is protected during transmission on the air interface by applying ARQ measures. As is the case for MAC, RLC is a sublayer of OSI layer 2.

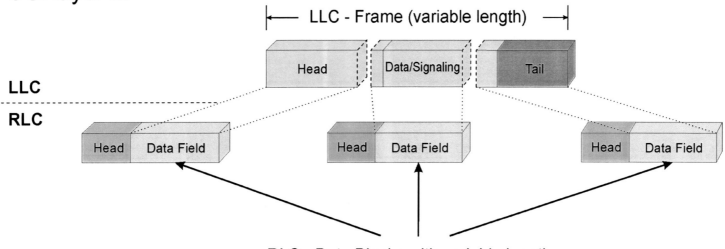

RLC - Data Blocks with variable length
dependent on coding scheme

The RLC / MAC Frame Structure:

The following four slides illustrate:

ö **The RLC/MAC Data Block in Downlink Direction.**

This format is used to transfer LLC-frames to the mobile station

ö **The RLC/MAC Data Block in Uplink Direction.**

This format is used to transfer LLC-frames from the mobile station to the network

ö **The RLC/MAC Control Block in Downlink Direction.**

This format is used to transfer RLC/MAC control messages on PACCH, PBCCH or PCCCH to the mobile station

ö **The RLC/MAC Control Block in Uplink Direction.**

This format is used to transfer RLC/MAC control messages on PACCH from the mobile station to the network

Format of RLC/MAC Data Blocks in Downlink Direction

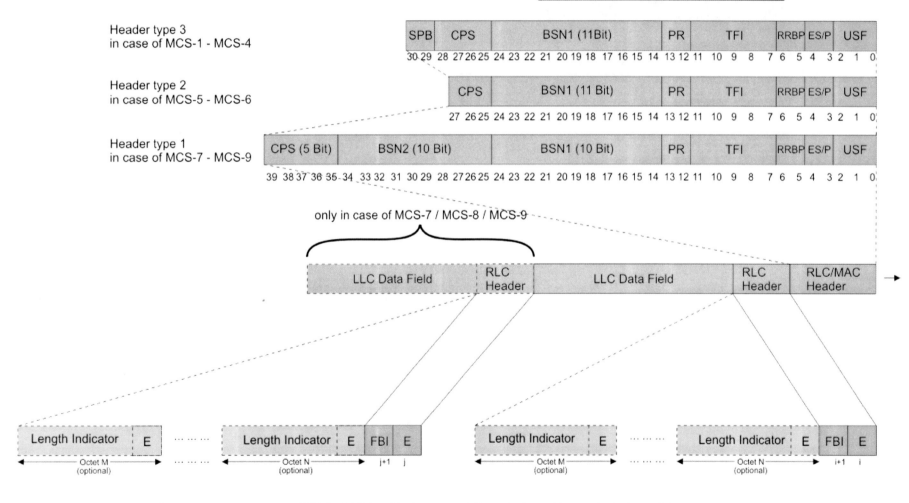

Header type 3
in case of MCS-1 - MCS-4

| SPB | CPS | BSN1 (11Bit) | PR | TFI | RRBP | ES/P | USF |

30 29 28 27 26 25 24 23 22 21 20 19 18 17 16 15 14 13 12 11 10 9 8 7 6 5 4 3 2 1 0

Header type 2
in case of MCS-5 - MCS-6

| CPS | BSN1 (11 Bit) | PR | TFI | RRBP | ES/P | USF |

27 26 25 24 23 22 21 20 19 18 17 16 15 14 13 12 11 10 9 8 7 6 5 4 3 2 1 0

Header type 1
in case of MCS-7 - MCS-9

| CPS (5 Bit) | BSN2 (10 Bit) | BSN1 (10 Bit) | PR | TFI | RRBP | ES/P | USF |

39 38 37 36 35 34 33 32 31 30 29 28 27 26 25 24 23 22 21 20 19 18 17 16 15 14 13 12 11 10 9 8 7 6 5 4 3 2 1 0

only in case of MCS-7 / MCS-8 / MCS-9

| LLC Data Field | RLC Header | LLC Data Field | RLC Header | RLC/MAC Header |

| Length Indicator | E | ... | Length Indicator | E | FBI | E | | Length Indicator | E | ... | Length Indicator | E | FBI | E |

Octet M (optional) ... Octet N (optional) j+1 j Octet M (optional) ... Octet N (optional) i+1 i

Format of RLC/MAC Data Blocks in Uplink Direction

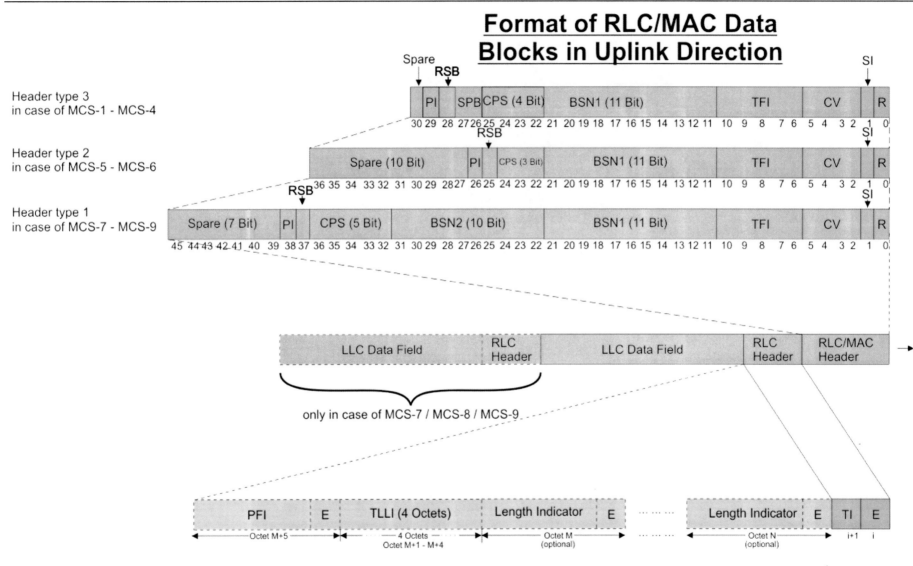

Dependency of LLC Data Field length on MCS:

LLC Data Field

≤ 22 Octets (depends on size of RLC Header)   <- in case of MCS-1

≤ 28 Octets (depends on size of RLC Header)   <- in case of MCS-2

≤ 37 Octets (depends on size of RLC Header)   <- in case of MCS-3

≤ 44 Octets (depends on size of RLC Header)   <- in case of MCS-4

≤ 56 Octets (depends on size of RLC Header)   <- in case of MCS-5 and MCS-7

≤ 68 Octets (depends on size of RLC Header)   <- in case of MCS-8

≤ 74 Octets (depends on size of RLC Header)   <- in case of MCS-6 and MCS-9

Format of RLC / MAC Control Blocks in Downlink Direction

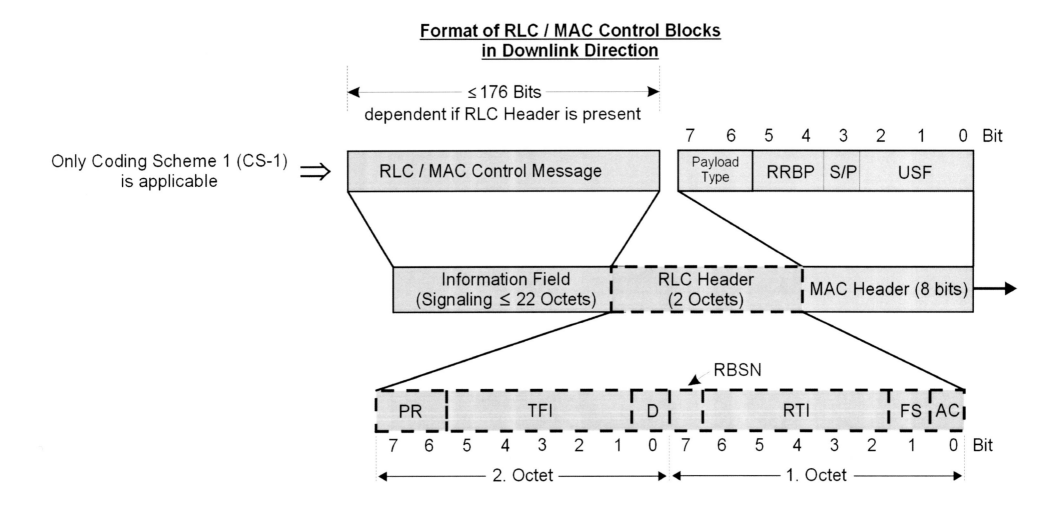

Only Coding Scheme 1 (CS-1) is applicable ⟹

≤ 176 Bits
dependent if RLC Header is present

RLC / MAC Control Message

7	6	5	4	3	2	1	0	Bit
Payload Type		RRBP		S/P		USF		

Information Field (Signaling ≤ 22 Octets) | RLC Header (2 Octets) | MAC Header (8 bits) →

RBSN

PR		TFI					D		RTI					FS	AC
7	6	5	4	3	2	1	0	7	6	5	4	3	2	1	0	Bit

← 2. Octet → ← 1. Octet →

Format of RLC / MAC Control Blocks in Uplink Direction

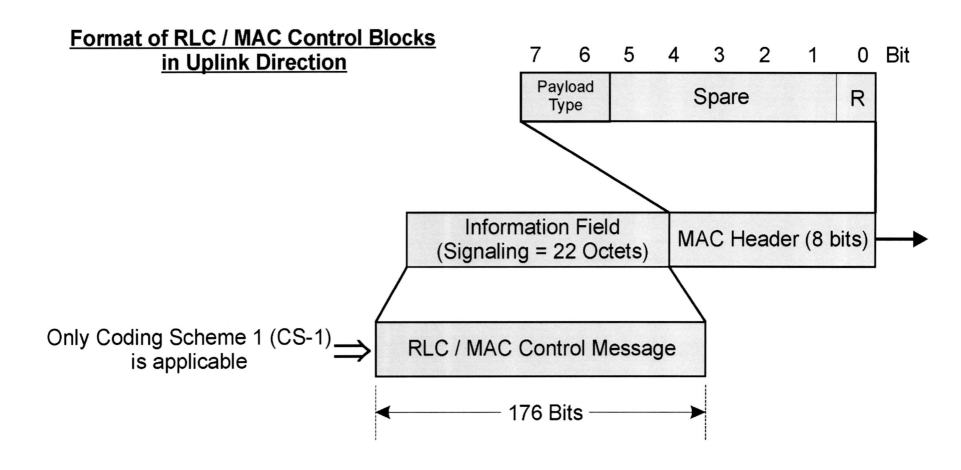

Only Coding Scheme 1 (CS-1) is applicable

(1) Explanation of RLC/MAC Block Information Elements:

Information Element	Frame Type	Explanation
USF	DL Control DL Data	Uplink State Flag / Details are provided in part 3
R-Bit	UL Control UL Data	The Re-try bit indicates to the network whether the mobile station needed to send its most recent initial access message (**CHAN_REQ** / **PACK_CHAN_REQ**) once or multiple times. By evaluating the R-bit from the various ongoing transactions, the system and/or the operator is able to determine the CCCH / PCCCH load. For instance, this is one criteria to decide if an(other) PCCCH needs to be allocated or not
SI-Bit	UL Data	The Stall Indicator bit is part of the MAC header in each RLC/MAC data and control block in uplink direction. By means of the SI-bit the mobile station indicates whether an acknowledgement from the network is required because the transmit window is exceeded (see ARQ). Obviously, the SI-bit is only meaningful in case of RLC acknowledged operation mode.
S/P-Bit	DL Control	The Supplementary / Polling bit validates the RRBP field. If the S/P bit is set to '1', the RRBP is valid, if S/P='0', the RRBP is not valid.

(2) Explanation of RLC/MAC Block Information Elements:

ES/P-Field	DL Data	The 2 bit long EGPRS Supplementary / Polling field validates the RRBP field and defines the content of the next uplink control block.
RRBP-Field	DL Control DL Data	RRBP is the abbreviation for Relative Reserved Block Period. The 2 bit long parameter allocates a single radio block in uplink direction to the addressed mobile station. Dependent on the actual coding of RRBP, the mobile station needs to wait a defined number of TDMA frames before it is allowed to transmit its control message on the same PDCH. This polling mechanism is used by the network during a downlink TBF to request either a **PACK_CTRL_ACK** message or a **PACK_DL_ACK** message from that mobile station.
CV-Field	UL Data	Abbreviation for Countdown Value. The 4 bit long CV indicates towards the network whether the mobile station intends to release an ongoing uplink TBF. More information about the countdown procedure can be obtained in part 3.
Payload Type	UL Control DL Control	The 2 bit long payload type field is part of each RLC/MAC control block in uplink and downlink direction. The payload type informs the receiver in downlink direction, if there is a RLC header in that RLC/MAC control block or not. In GPRS the payload type field was use to discriminate between RLC data blocks and RLC control blocks.

(3) Explanation of RLC/MAC Block Information Elements:

FBI-Bit	DL Data	The FBI or Final Block Indicator bit is a mandatory part of the RLC header of each downlink RLC/MAC data block. When set, it indicates that this RLC/MAC-block is the last downlink RLC data block of a TBF to be sent on a PDCH. The network intends to release the downlink TBF.
CPS-Field	DL Data UL Data	The Coding and Puncturing Scheme indicator field in EPRS RLC/MAC headers is used to indicate the channel coding and the puncturing scheme used for the following data blocks.
SPB-Field	UL Data DL Data	The 2 bit long Split Block indicator field is used in headertype 3 to indicate whether the user data is retransmitted using block resegmentation or not. (-> Acknowledged Mode for RLC/MAC operation)
TI-Bit	UL Data	The TLLI Indicator bit is set when the RLC/MAC block contains an optional TLLI in its RLC header.
AC-Bit	DL Control	The Address Control-bit indicates that the second octet of the RLC header with the TFI information is present, too.
FS-Bit	DL Control	The Final Segment bit is part of the optional RLC-header within downlink RLC/MAC control blocks. It indicates to the mobile station that the current RLC/MAC control block is the final segment of a multi-segment RLC/MAC control message. See also RBSN, RTI.

(4) Explanation of RLC/MAC Block Information Elements:

RTI-Field	DL Control	The 5 bit long Radio Transaction Identifier field is part of the optional RLC header within downlink RLC/MAC control blocks. Since there may be more than one simultaneous control transaction ongoing at any given time, the RTI-field is required to distinguish among these ongoing transactions. Each transaction is assigned its unique RTI value (see also RBSN and FS).
D-Bit	DL Control	The Direction-bit is part of the optional RLC header of downlink RLC/MAC control blocks. It indicates whether the following TFI-field identifies an uplink or downlink TBF. This is necessary because RLC/MAC control blocks are sent on the bi-directional PACCH as opposed to RLC/MAC data blocks which are sent on PDTCH). Since there may be identical TFIs in use in uplink and downlink direction, not necessarily for the same mobile station, the D-bit helps to avoid ambiguities.

(5) Explanation of RLC/MAC Block Information Elements:

TFI-Field	DL Data UL Data DL Control	The TFI (Temporary Flow Identity) is part of the RLC/MAC-header of each data block, being sent by the network. In downlink RLC/MAC control messages, the presence of the TFI is optional. The TFI is also present in all RLC/MAC-data blocks, being sent by the mobile station. In uplink MAC/RLC-control blocks, the TFI is not part of the header. Identification is taken care of by the message being transferred. Note that identical TFIs may be assigned to two different but simultaneous uplink and downlink transactions. That is, a TFI may be used for a TBF in uplink direction while at the same time, a downlink TBF is using the same TFI. Of course, in each direction the TFIs are unambiguous at any given moment. The lifetime of a TFI is limited and depends on the lifetime of its TBF. With the release of a TBF, the related TFI becomes available again and can be assigned to another transaction.
PR-Field	DL Data DL Control	The PR or Power Reduction field is a mandatory part of the RLC header in each downlink RLC/MAC data block and optional in the downlink RLC/MAC control block. Dependent on its coding it informs the addressed mobile station about the power level of that PDCH relative to the BCCH power level.

(6) Explanation of RLC/MAC Block Information Elements:

E-Bit	DL Data UL Data	The Extension-Bit is part of the, mandatory and optional, RLC-Header within each RLC/MAC data block in uplink and downlink direction. The E-Bit indicates whether there is a following optional octet after the present one, or not. (-> Example)
LI-Field	DL Data UL Data	The 7-bit long Length Indicator Field is part of the optional octets of the RLC header of RLC/MAC data blocks .It is used to delimit the length of segments of different LLC frames being part of one RLC/MAC data block. **Only the last segment of any LLC PDU of a TBF has to be identified with a Length Identificator within the corresponding RLC data block.** (-> Example) At first sight it is surprising that the LI field per definition can take on values only between '0' and '74'dec and the value '127'dec though it could range from '0' to '127'dec. The answer to this question is given by the fact that even with MCS-9 the maximum amount of data within one RLC/MAC block is limited to 74 octets.

Example of LI-Field and E-Bit:

Only the last segment of any LLC PDU of a TBF must be identified with a Length Indicator within the corresponding RLC data block.

Therefore, the first segment of LLC PDU 3 is not delimited with an LI-Field.

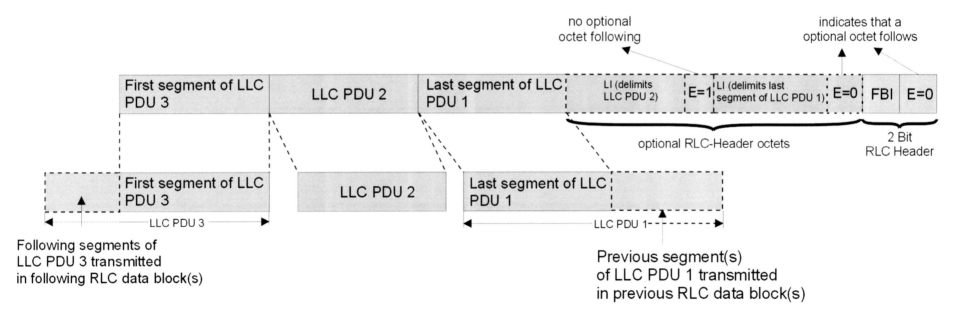

(7) Explanation of RLC/MAC Block Information Elements:

RBSN-Bit	DL-Control	The Reduced Block Sequence Number bit is part of the optional RLC header within downlink RLC/MAC control blocks. Obviously, RBSN can take on the binary values '0' and '1'. RBSN is used for error protection of transferred control messages. Its value corresponds to the internal V(CS) counter and toggles between '0' and '1' for each control block that is sent within a transaction. This transaction may involve multiple control messages. There is some similarity between the RBSN and the SSN, used on the GSM air interface. An example for the use of the RBSN is provided in the figure on the right hand side. Note that there may be more than one simultaneous control transaction ongoing at a given time, each one uniquely identified by the RTI. The final segment is identified by the FS-bit set to '1'.

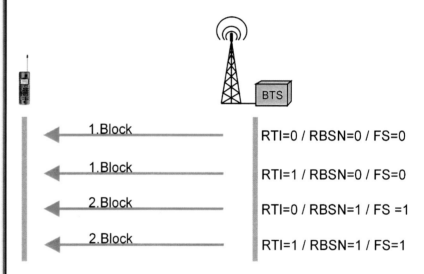

(8) Explanations for RLC/MAC Block Information Elements:

RSB-Bit	UL Data	The Resent Block Bit indicates whether all of the RLC data blocks contained within the EGPRS radio block are sent for the first time (bit=0) or at least one RLC data block has been transmitted before (bit=1). Note: The use of this bit will be reconsidered in future versions of the specification.
PI-Bit	UL Data	The Packet Flow Identifier Indicator bit indicates the presence of the optional PFI field.
PFI-Field	UL Data	The Packet Flow Identifier (PFI) is used to identify a Packet Flow Context (PFC). It defines the QoS (Quality of Service) that must be provided by the BSS for a given packet flow between the MS and the SGSN, i.e., for the Um and Gb interfaces combined.

(1) Logical Link Control (LLC):

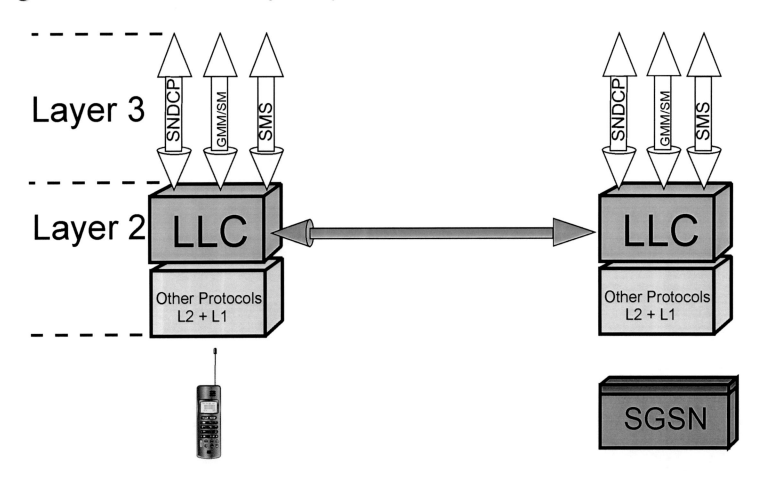

(2) Logical Link Control (LLC):

- The LLC protocol is the lowest EGPRS protocol that is independent from the air interface protocols which are used. This is most important, making the EGPRS core network design as independent as possible of the air interface.

- The LLC provides services to the network layer protocols that EGPRS uses. These are, in particular, the SNDCP (ì   User Data Transfer), SMS and GMM / SM (GPRS Mobility Management / Session Management).

Functions of LLC:

Since LLC is considered a part of OSI layer 2, most of its functions relate to typical data link layer tasks. Similarly to RLC, LLC offers acknowledged and unacknowledged operation modes. It is unusual that LLC is also in charge of the EGPRS ciphering procedure:

ö   Encapsulation of higher layer protocol data units into LLC data units. This particularly applies to data units from SNDCP which are tailored to fit into one LLC data unit.

ö   Delivery of data units to the higher layer in the *correct sequence*.

ö   Provision of different service levels. Unacknowledged and acknowledged operation modes especially are supported by LLC.

ö   Establishment of virtual connections in acknowledged operation mode.

ö   Ciphering and de-ciphering of the information fields that are included.

(1) The LLC Operation Modes:

LLC can offer both acknowledged and unacknowledged operation modes.

ö **Acknowledged Operation:**
The acknowledged operation mode is based on the ABM procedure, as are LAPD and other protocols. It is important not to become confused by the fact that ABM involves the establishment of a virtual connection between the two peers. Initially, this may seem strange with regard to a packet radio service. However, since the connection is virtual, there is no conflict between packet switching and the use of the ABM-procedure for LLC.

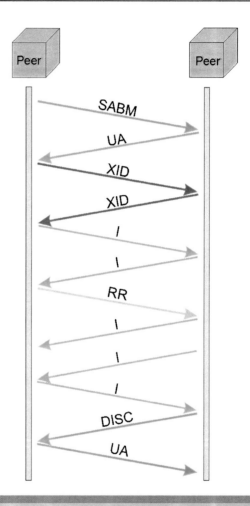

Note: the acknowledged operation mode is only allowed for the transfer of user data (⇔ SNDCP) but it must not be used for GMM, SM or SMS procedures.

239

(2) The LLC Operation Modes:

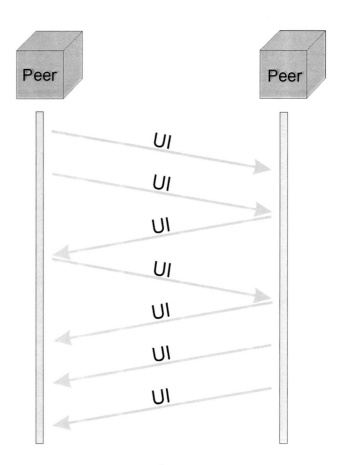

ö **Unacknowledged Operation:**
The unacknowledged operation mode is based on the ADM procedure, which is known in connection with LAPD. The application of ADM is mandatory for GMM, SM and SMS. Thus, procedures such as PDP context activation or EGPRS Attach are performed in ADM.

Note: the unacknowledged operation mode may also be used for data transfer. This depends on the level of service (QoS) of the respective transaction.

240

(1) The LLC Frame Format / I+S- , S- and UI-Frames:

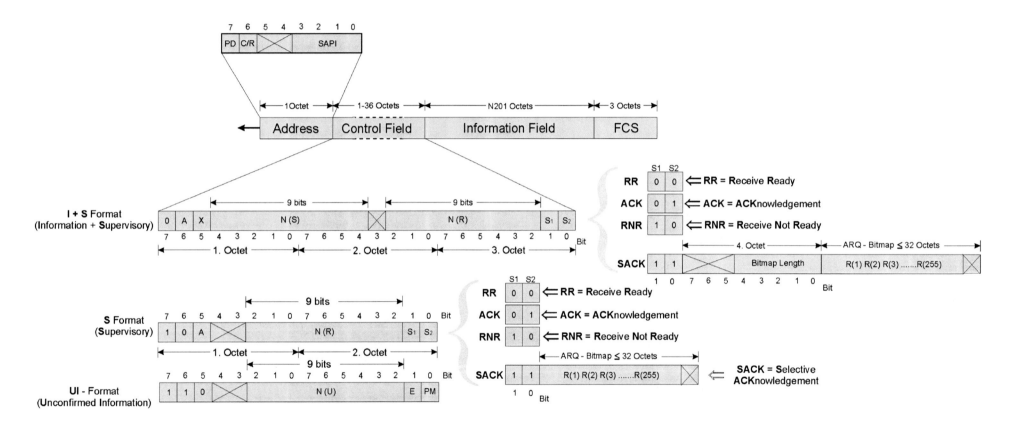

241

(2) The LLC Frame Format / U-Frames:

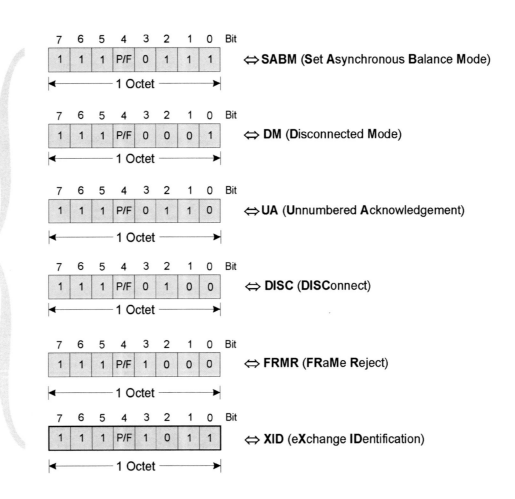

242

(1) Explanations for the LLC Frame Format :

The LLC frame format is based on HDLC though the usual start and end flags are missing. Those who are familiar with the LAPD protocol will see many similarities between the formats of LAPD and LLC. However, there are *essential differences*:

ö   LLC does not have an exclusive information frame. LLC introduces the **I+S** format, that is, each piece of information also carries supervisory information. Obviously, this measure helps to save resources. The maximum length of the information field of an **I+S**-frame depends on N201 that may be negotiated prior to transmission between the two peers.

ö   Like RLC, LLC applies ARQ for error recognition and correction. Thus, the supervisory and information frame types need to be modified to suit the requirements of ARQ. For instance, both frames must be able to convey an entire bit map to identify information frames which have been received incorrectly. Accordingly, LLC introduces the **SACK**-frame (SACK ì Selective ACKnowledgement).

(2) Explanations for the LLC Frame Format :

ö   The **I+S**- and the **S**-format frames contain a new parameter, the A-bit. The A-bit is used to invoke acknowledgement from the peer.

ö   In LAPD, **UI** means unnumbered information. In LLC, **UI** means *unconfirmed information*. The difference is that LLC numbers **UI**-frames but, as is the case in LAPD, there is no means to request the re-transmission of a **UI**-frame. Another difference is that LLC **UI**-frames offer means for error recognition, but do not offer means for error correction. However, numbering enables LLC to detect and discard frames that are received twice. As is the case for the **I+S**-frame, the maximum number of octets within an **UI**- frame depends on N201.

ö   In LLC unacknowledged operation mode, one needs to distinguish between protected and unprotected mode. In the protected mode, the check sum works for the entire frame, including the header and the information field. In the unprotected mode, the check sum only protects the header and the first four octets of the information field. These four octets represent the maximum size of the header of the embedded SNDCP data unit. Thus, even in LLC unprotected mode, transmission errors in the SNDCP header are recognized. The Protection Mode is determined by the PM bit.

(1) Ciphering in EGPRS :

ö   Unlike the procedure in pure GSM, ciphering is carried out between the SGSN and the mobile station in EGPRS.

ö   Ciphering is one of the tasks of LLC.

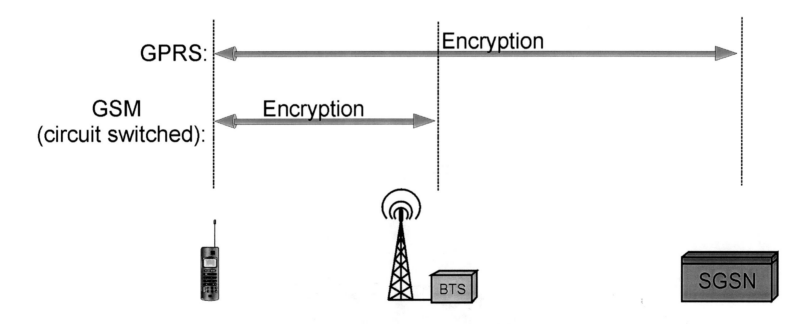

245

(2) Ciphering in EGPRS :

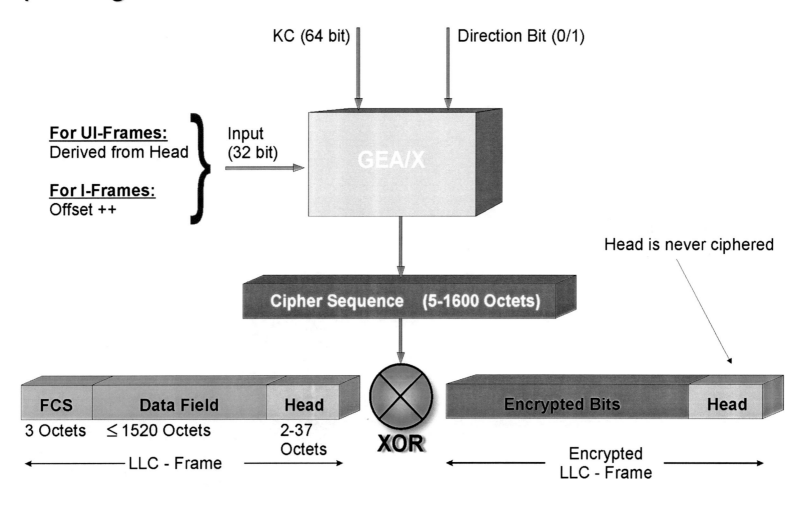

The SubNetwork Dependent Convergence Protocol (SNDCP):

ö As for the LLC protocol which has already been described, the SNDCP ranges from the mobile station to the SGSN.

ö The SNDCP mainly relies on the services of the LLC.
This applies in particularly to *in-sequence delivery of SNDCP data units, SN-PDUs and to the actual transmission between the mobile station and the SGSN.*

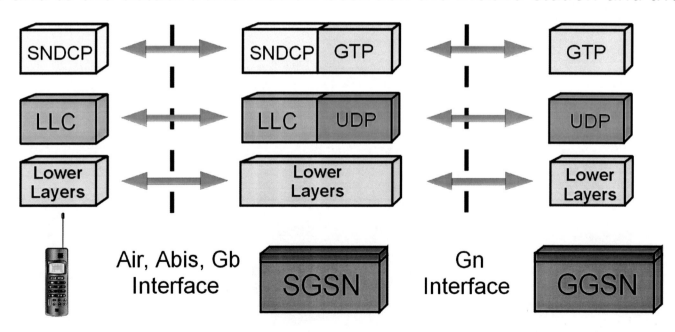

247

(1) Functions of SNDCP:

ö   Interfacing the EGPRS protocol stack between the SGSN and the mobile station to the supported packet data protocols (PDP) which are IP, PPP and IHOSS.
Note: SNDCP will be relayed to the GPRS tunneling protocol (GTP) within the SGSN.

ö   Compression and decompression of user data.

ö   Compression and decompression of packet headers. Note that this function is only applicable to TCP/IP but does not apply to e.g. UDP/IP.

ö   Segmentation and de-segmentation of so called N-PDUs into SN-PDUs. N-PDUs are the data packets of the application protocols.

ö   SN-PDUs are forwarded after optional compression to LLC for transmission within LLC **UI**- or **I+S**-frames. **UI**-frames are used for the unacknowledged operation mode and **I+S**-frames are used in case of the acknowledged operation mode. Note that each SN-PDU is embedded into a single **I+S**- or **UI**-frame.

(2) Functions of SNDCP:

ö   If a change of SGSN occurs during an ongoing packet transfer, SNDCP
should forward all unacknowledged SN-PDUs to the new SGSN for
downlink transmission.
(This only works for the acknowledged operation mode)

ö   For actual transmission, SNDCP relies completely on LLC. This also
applies to the in-sequence delivery of SN-PDUs.

ö   SNDCP selects the LLC transmission mode, acknowledged or
unacknowledged. The choice of LLC transmission mode depends
on the negotiated QoS profile.

(3) Functions of SNDCP:

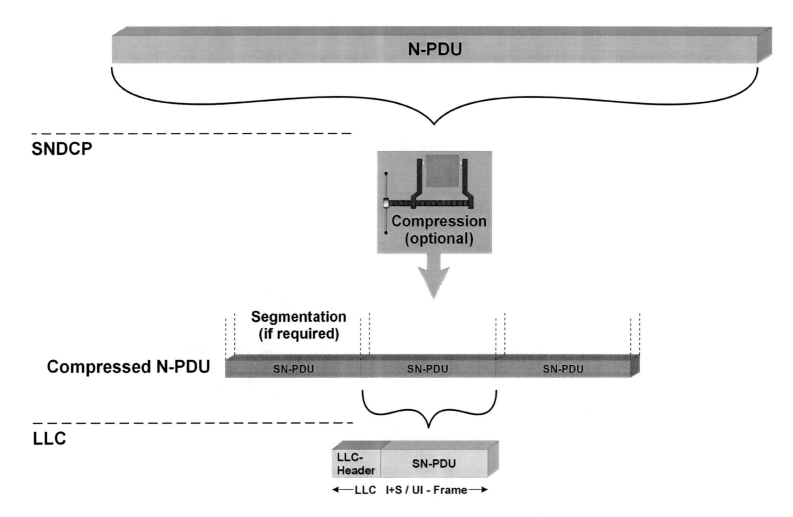

(1) Compression Techniques of SNDCP:

ö   Header compression is based on RFC 1144. It is only applicable for TCP/IP and cannot be used for other data units. RFC 1144 makes use of extensive redundancy within the TCP/IP header when a transaction is established   .

ö   Data Compression is based on the ITU-T V.42bis recommendation (see part 2). It is applicable to all supported network protocols, these being IP, PPP and IHOSS.

ö   Note that the data compression process is applied not only to the data field itself but *also to the header which may well have already undergone compression*. Thus, compression is carried in two steps, for TCP/IP headers at least.

ö   In a later phase of GPRS\EGPRS, additional compression schemes may be provided by the SNDCP. Since there is an optional negotiation phase for the used compression algorithms and their parameters, there will not be a conflict with previous implementations of SNDCP.

(2) Compression Techniques of SNDCP (⇔ TCP/IP):

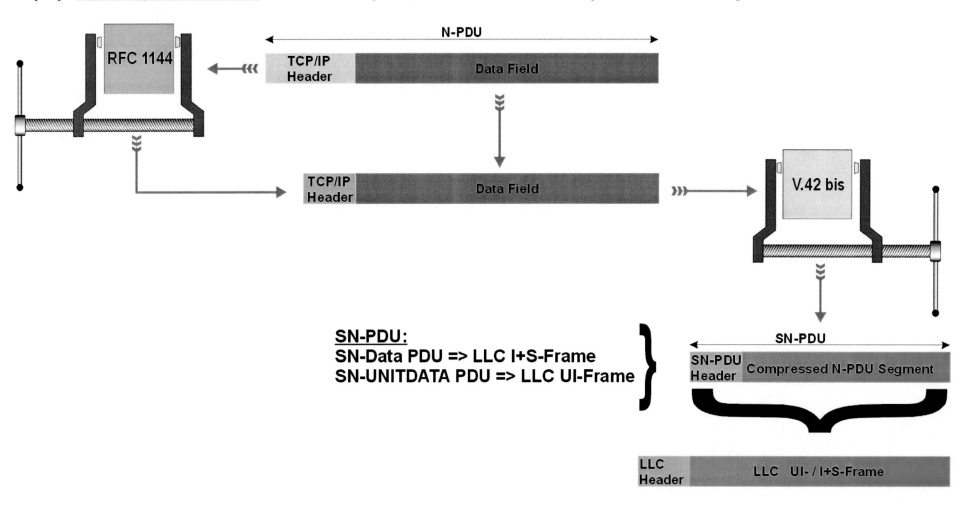

The SN-PDU Format

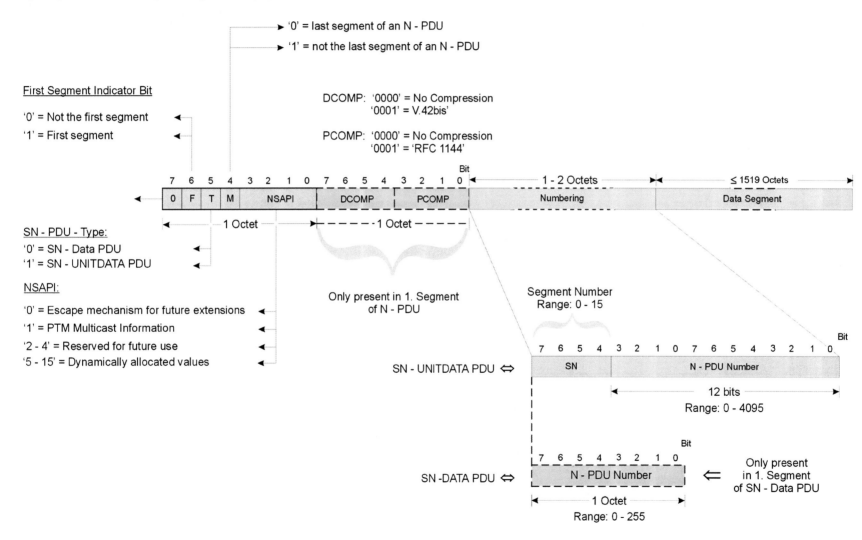

'0' = last segment of an N - PDU

'1' = not the last segment of an N - PDU

First Segment Indicator Bit

'0' = Not the first segment

'1' = First segment

DCOMP: '0000' = No Compression
'0001' = V.42bis'

PCOMP: '0000' = No Compression
'0001' = 'RFC 1144'

| 7 | 6 | 5 | 4 | 3 | 2 | 1 | 0 | 7 | 6 | 5 | 4 | 3 | 2 | 1 | 0 | 1 - 2 Octets | ≤ 1519 Octets |

| 0 | F | T | M | NSAPI | DCOMP | PCOMP | Numbering | Data Segment |

1 Octet — — — 1 Octet — — —

SN - PDU - Type:

'0' = SN - Data PDU

'1' = SN - UNITDATA PDU

NSAPI:

'0' = Escape mechanism for future extensions

'1' = PTM Multicast Information

'2 - 4' = Reserved for future use

'5 - 15' = Dynamically allocated values

Only present in 1. Segment
of N - PDU

Segment Number
Range: 0 - 15

| 7 | 6 | 5 | 4 | 3 | 2 | 1 | 0 | 7 | 6 | 5 | 4 | 3 | 2 | 1 | 0 |

SN - UNITDATA PDU ⇔ | SN | N - PDU Number |

12 bits
Range: 0 - 4095

| 7 | 6 | 5 | 4 | 3 | 2 | 1 | 0 |

SN -DATA PDU ⇔ | N - PDU Number |

Only present
in 1. Segment
of SN - Data PDU

1 Octet
Range: 0 - 255

GPRS Mobility Management (GMM):

ö   As in circuit switched GSM, GPRS\EGPRS Mobility Management ensures that a subscriber is able to roam.

ö   GPRS\EGPRS provides roaming within and outside the H-PLMN.

ö   For GPRS\EGPRS, roaming is basically a function, residing in and controlled by the SGSN and the mobile station.

ö   GPRS\EGPRS introduces a number of new identifiers within the network and the mobile station that shall be presented on the following pages.

(1) GMM Identifiers:

ö ## The P-TMSI:

The Packet Temporary Mobile Station Identity uniquely identifies a given mobile station within an SGSN area. The P-TMSI has a fixed length of 4 octets and is dynamically assigned by the SGSN upon network access or change of the location. As in circuit switched TMSI, the P-TMSI is introduced for improved confidentiality.

ö ## The Routing Area Identification (RAI)

Similarly to the related Location Area Identification (LAI) in circuit switched GSM, the RAI uniquely identifies a group of one or more adjacent cells that cover a specific geographic area. The RAI consists of the LAI plus the 1 octet long Routing Area Code (RAC). Therefore, one or more routing areas form one location area.

(2) GMM Identifiers:

The Temporary Logical Link Identifier (TLLI):

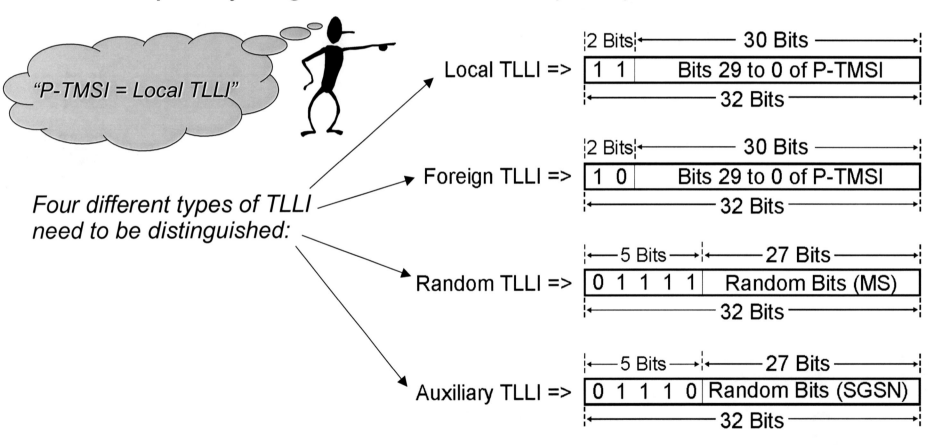

"P-TMSI = Local TLLI"

Four different types of TLLI need to be distinguished:

Local TLLI =>

2 Bits	30 Bits
1 1	Bits 29 to 0 of P-TMSI

32 Bits

Foreign TLLI =>

2 Bits	30 Bits
1 0	Bits 29 to 0 of P-TMSI

32 Bits

Random TLLI =>

5 Bits	27 Bits
0 1 1 1 1	Random Bits (MS)

32 Bits

Auxiliary TLLI =>

5 Bits	27 Bits
0 1 1 1 0	Random Bits (SGSN)

32 Bits

The Four Different TLLIs:

ö **The Local TLLI**
The local TLLI is in fact the P-TMSI. It can only be used by the mobile station in the routing area in which the P-TMSI was allocated.

ö **The Foreign TLLI**
Similarly to the local TLLI, the foreign TLLI is built up from the P-TMSI. To distinguish between a foreign TLLI and a local TLLI, the two MSB's of a foreign TLLI are set to '10'$_{bin}$. A foreign TLLI is used by the mobile station for identification purposes, if the P-TMSI used was allocated in another routing area.

ö **The Random TLLI**
The random TLLI is used by the mobile station to establish a logical link if no valid P-TMSI is available or, when the mobile station intends to originate anonymous access. Note that as opposed to local and foreign TLLI, bit 31 to 27 of a random TLLI have fixed codes.

ö **The Auxiliary TLLI**
The auxiliary TLLI can only be allocated by the SGSN to continue with the establishment of anonymous access. Similarly to the random TLLI, bits 31 to 27 have fixed codes.

Important GMM Procedures:

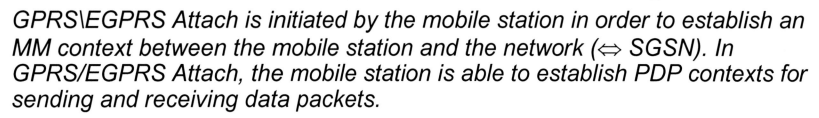

ö **GPRS\EGPRS Attach / Detach:**

GPRS\EGPRS Attach is initiated by the mobile station in order to establish an MM context between the mobile station and the network (⇔ SGSN). In GPRS/EGPRS Attach, the mobile station is able to establish PDP contexts for sending and receiving data packets.

ö **Important Content of a GMM-Context:**

Authentication Triplets: (RAND, SRES, Kc): received by the SGSN from the subscriber 's HLR.

CKSN / Kc: *Kc / CKSN pair currently in use.*

DRX parameters *if DRX is used.*

Identifier: *P-TIMSI and P-TIMSI signature for identification of the GMM context.*

The GMM context can be updated at cell or routing area update.

The GMM context is stored in the SGSN.

Important GMM States in the SGSN and the Mobile Station:

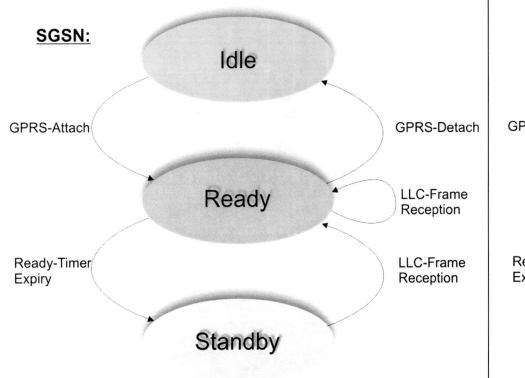

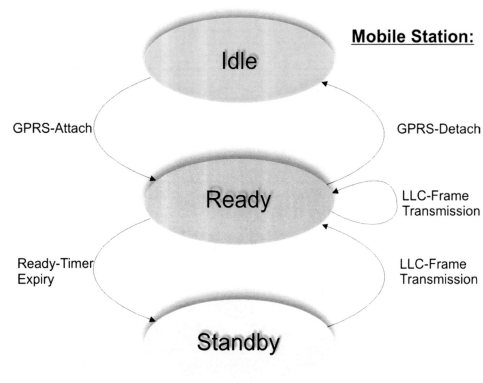

Important GMM Procedures:

ö   ## Routing Area Update (Intra-SGSN / Inter-SGSN)

A Routing Area Update is performed by the mobile station when it detects that the serving cell belongs to another routing area (⇔ (PACK)_SYS_INFOs).

Note that the mobile station performs a cell update as long as the READY timer (T 3314 / Default = 44 s) has not expired. The READY timer is restarted each time the mobile station issues an LLC frame. This timer is also applicable to the SGSN. A cell update consists of sending an LLC frame to the SGSN. Upon reception of the LLC frame, the READY timer is also restarted in the SGSN. Of course, when passing a routing area border, the mobile station is still required to perform a routing area update rather than a cell update.

EGPRS Session Management (SM):

ö   In GPRS\EGPRS, a session involves the transfer of application data from and/or to a mobile station.

ö   Before the actual data transmission may start, session management involves a handshaking procedure between the mobile station, the SGSN and the GGSN.

ö   Most importantly, handshaking establishes a PDP context between the mobile station, the SGSN and the GGSN that includes, amongst others, the negotiated QoS profile and the respective PDP addresses.

ö   GPRS session management also introduces so-called Anonymous Access. That is, when the subscriber does not need to identify himself to the SGSN/GGSN. Identification is carried out towards the application peer on a higher layer.

ö   Anonymous access is only possible for specific applications such as for *toll road services*.

ö   Anonymous access is not supported by GPRS Rel. 99.

Important parameters / content of a PDP context:

NSAPI: Network Service Access Point Identifier to identify the PDP context on the SNDCP layer; selected by the MS during PDP context activation.

SAPI: Service Access Point Identifier on LLC layer

PDP address, type and organization:

PDP address to be used by the MS e.g. dynamic IP address, and the PDP type e.g. IP or PPP.

GGSN address: for cell update that involves the change of the SGSN.

QoS: subscribed, requested, allocated: three different sets of QoS profiles.

An MS may support multiple simultaneous sessions and activated PDP context.

A PDP context may be changed. This can be initiated by the MS, the SGSN or the GGSN.

Message Format for GMM and SM:

Note that SM and GMM messages are transferred transparently between the mobile station and the SGSN.

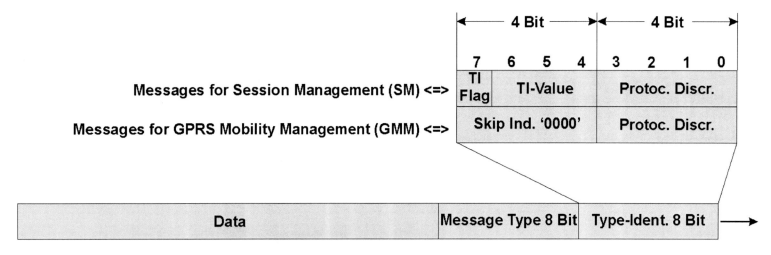

GMM => PD = '1000' = '8'

SM => PD = '1010' = 'A'

EDGE from A - Z

(1) Summarizing the Generic Part:

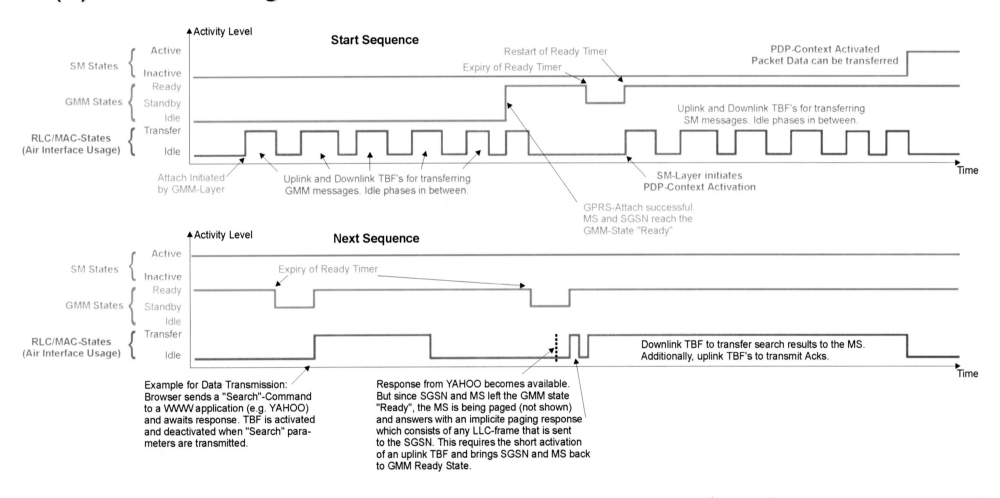

(2) Summarizing the Generic Part:

The state diagram on the previous slide highlights the following characteristics of GPRS\EGPRS :

ö   Real "Resource on Demand" concept. TBF's are only active when packet data needs to be transmitted. In between times, the resources are idle (please compare with the concept of e.g. SDCCH in GSM)

ö   Before the mobile station and the SGSN have activated a GMM context, a PDP context cannot be activated between the mobile station, SGSN and GGSN.

ö   Before a PDP-context is activated, packet data cannot be transmitted.

ö   Page Response in EGPRS is carried out implicitly. GPRS\EGPRS does not have a page response message. When the network needs to page the mobile station, that is when the SGSN and the mobile station are not in the GMM state "READY", the mobile station responds with any LLC frame that is sent to the SGSN.

Comparison between

GPRS ⇔ EDGE ⇔ UMTS

EDGE: standardization and evolution

EDGE started up in 1997 as a technological enhancement to the GSM system (ETSI) enabling higher data throughput rates to be attained:

    HSCSD (57.6kbit/s) ö   ECSD (230.4kbit/s, 8TS)

    GPRS (172kbit/s)   ö   EGPRS (491kbit/s, 8TS)

EDGE was adopted in 1998 for ANSI IS-136HS radio interface by the UWCC (Universal Wireless Convergence Consortium) to enable global roaming for GSM and IS-136.

EDGE phase 2 standardization is directed towards improvement in multimedia and real-time services (Rel. 2000).

Interfacing the GSM / GPRS / EDGE base station subsystem (BSS) to the future UMTS core network brings us to the concept of GERAN (GSM / EDGE Radio Access Network).

GERAN (GSM / EDGE Radio Access Network):

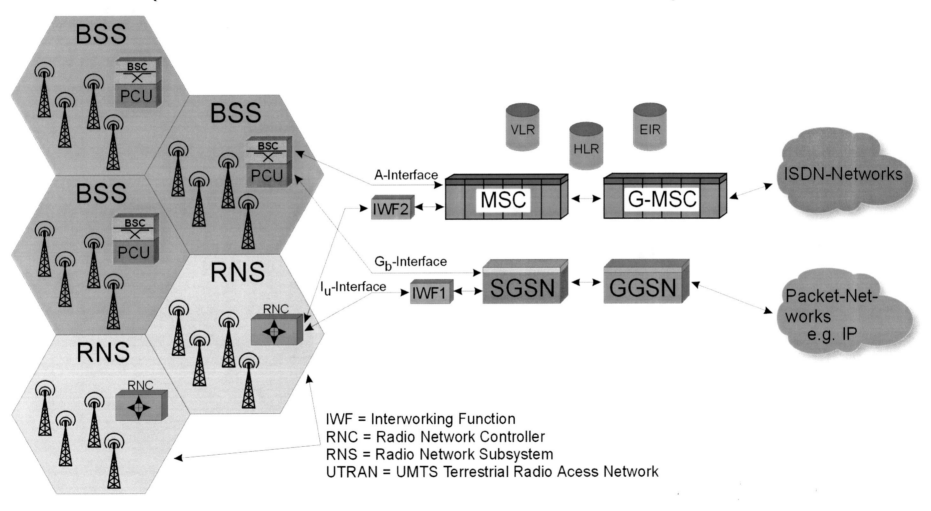

IWF = Interworking Function
RNC = Radio Network Controller
RNS = Radio Network Subsystem
UTRAN = UMTS Terrestrial Radio Acess Network

268

Introduction Scenarios:

Using UMTS in high traffic urban areas whilst offering UMTS-like services on the basis of EGPRS in low traffic rural areas, will cut investment costs.

Support of EDGE for IS-136 by US operators might secure the development of EDGE products.

A delay in WCDMA products while data services take off using GPRS would completely change the situation.

<u>Many unpredictable pro´s and con´s:</u>

Will WCDMA products be available in time?

Mixed introduction of UMTS and GERAN will require triple (or even quadro mode (GSM) / GPRS / EDGE / WCDMA terminals. Will the terminals be available?

Will the US situation change after the AT&T decision for GSM?

Expected feasibility of mobile communication systems:

	GSM	HSCSD	GPRS	ECSD	EGPRS	UMTS FDD	UMTS TDD
max. bitrate DL (kbit/s)	9,6	76,8	160	473,6	473,6	384[1]	2048[2]
estimated average bitrate (kbit/s)	9,6	38	60	170	170	240[1]	1024[2]
max. users / cell	ca. 45	ca. 11	ca. 11	ca. 11	ca. 11	ca. 9	1
Infrastructure	existing HW	extension	extension	extension	extension	new	new
new cell locations	-	no	no	rate dependent	rate dependent	+	+
new hardware	-	few	few	+	+	++	++
new software	-	+	+	+	+	++	++
Implementation (year)	1991	2000	2000	-	2002	2002	2004

1) moving subscriber
2) stationary subscriber

Taken from "Turbolader für Funk-Bits"
ct-magazin für computer technik Nr. 19 / 2000